Disaster Management

Disaster Management

Preparedness and Mitigation

Saima Akbar
Suraj Kumar Singh
Shruti Kanga

ISTE

WILEY

First published 2024 in Great Britain and the United States by ISTE Ltd and John Wiley & Sons, Inc.

ISTE Ltd
27-37 St George's Road
London SW19 4EU
UK

www.iste.co.uk

John Wiley & Sons, Inc.
111 River Street
Hoboken, NJ 07030
USA

www.wiley.com

Library of Congress Control Number: 2024939223

British Library Cataloguing-in-Publication Data
A CIP record for this book is available from the British Library
ISBN 978-1-78630-996-9

Contents

Preface

Over time, anthropogenic activities for commercialization and livelihood needs have placed an immense amount of stress on the planet, which has increased our vulnerabilities on a large scale. The only solution to these disasters lies in awareness, preparedness and disaster risk reduction, which can be attained through scientific means.

Through the spatial understanding of natural- and human-induced hazards, policymakers can make vigilant decisions to minimize the harmful effects of natural disasters such as earthquakes, landslides, floods, droughts, volcanic eruptions and anthropogenic disasters such as nuclear, industrial and environmental. The management strategies such as activities, awareness campaigns, proactive initiatives and use of early warning systems can play an important role in the minimization of its impact on people and their properties.

The world has a physical and economic dilemma for humans since it is prone to several natural and man-made disasters each year. It will be tough to handle the numerous obstacles that arise after a disaster if no preparation is made. After critically analyzing different types of disasters and vulnerabilities, the book concludes that complete prevention of natural and anthropogenic disasters is not possible, however, we can reduce their effects through disaster risk reduction and preparedness measures, minimizing the damage and destruction caused by the disasters in India with a long-term perspective. These are effective tools for the management of disasters, which are discussed in a lucid way throughout the book.

About the book

Disaster Management: Preparedness and Mitigation is a comprehensive book focusing on managing disasters using effective preparedness and mitigation strategies. This book delves into essential principles and practices crucial for navigating through crisis, focusing specifically on proactive planning strategies and implementation of measures aimed at minimizing the impact of disasters on communities, infrastructure and the environment.

To facilitate a clearer understanding, the book has been divided into various chapters, each with an introduction giving details about disasters, hazards, classification of disasters, disaster management cycle and preparedness and mitigation measures to be taken into account in order to reduce the harmful impact of different disasters. The book also critically analyzes the vulnerability and recent case studies in India to various forms of disasters. Details regarding concepts of national disaster response forces are given, as well as safety tips during disasters, and details of natural, anthropogenic and environmental challenges such as climate change, global warming, as well as their risk reduction strategies.

A special attempt has been made for veterinarian, UG and PG students to understand in detail the management of different types disasters and disaster risk reduction strategies in preparedness, response and emergency phases. The book also discusses various factors influencing the vulnerability of people and society to the natural calamities and provides disaster risk reduction strategies to tackle disasters and identify the most suitable approaches for a particular hazardous situation. The case studies of India have been also discussed for the various preparedness and mitigation measures as a solution for various types of disaster provides preparedness and mitigation measures in order to help manage various types of disasters using sustainable approaches. It outlines what, how and when to take certain actions prior to, during and after a disaster strikes, providing basic information for managing disasters while highlighting the effects of disasters on life and the economy. The primary goal of writing this book is to equip the students by providing fundamental and general knowledge necessary in a disaster situation. This book contains chapters to help the students, researchers and policymakers develop complete understanding of disasters and provide effective preparedness and mitigation measures to reduce disaster risk. The book walks readers through every step of putting the initiative to promote a preparation, response and mitigation into practice. This book will be helpful for students, educators, community members, disaster managers, social workers, media persons and citizen. At the end of every chapter, descriptive questions and solved MCQs have been mentioned that will be beneficial for the students who are pursuing a degree or appearing for any competitive examination.

June 2024

protection by being informed of the different kinds of disasters. Disasters are events that cause widespread destruction, loss of life and disruption of the normal life system. Understanding the concept and root causes of disasters, we can together work toward preventing and effectively responding to these catastrophic events by reducing their impact on vulnerable populations to mitigate their impact. While some disasters are unavoidable, many can be prevented or mitigated through effective planning, early warning systems and community preparedness. By understanding the risks and vulnerabilities in a specific area, we can take steps to minimize the impact of disasters and protect the people at risk; therefore, by working together at different levels to support one another, we can better navigate the challenges presented by these disasters.

1.2. Classification of disasters

Disasters are sudden events that strike at any time and can have devastating effects on communities. It is important to understand the different types of disasters and how they are classified to better prepare and effectively respond to them to mitigate their impact. Disasters can be classified into several categories based on their cause and impact. These categories include natural disasters, man-made disasters, technological disasters and environmental disasters.

1) *Natural disasters*: these are the disasters that are caused by natural activities (processes) or natural phenomena, which means they are under the control of nature. These disasters are caused by natural forces, for example, droughts, tsunamis, floods, hurricanes, wildfires volcanic activities and earthquakes are exclusively of natural origin. These disasters can cause widespread damage to property and loss of life.

2) *Human-induced disasters*: these are the events that are triggered by human activities or result from human negligence or error. They are also called man-induced or anthropogenic disasters and are associated with human actions, such as industrial accidents, dam failure terrorist attacks, explosions, leakage of toxic or poisonous substances and oil spills. These disasters can also have serious impacts on humans and require immediate action to mitigate their impact. Human interaction with the environment has also resulted in some of the most major accidents.

1.2.1. *Other categories of disasters*

1) *Minor disaster*: a minor disaster is an event that causes limited damage and has a relatively low impact on the affected population, for example, small-scale localized incidents such as a minor fire, a small-scale industrial accident or a localized flood that affects a small community.

2) *Moderate disaster*: a moderate disaster is an event that causes a higher level of damage and has a more significant impact on the affected population, compared to a minor disaster. These may include larger scale incidents such as a moderate earthquake, a significant industrial accident affecting a larger area or a flood that affects several communities.

3) *Major disaster*: a major disaster is an event that results in extensive damage, poses a severe threat to human life and requires a substantial and coordinated response from local, regional or national authorities. These may include large-scale natural disasters such as tsunamis and earthquakes, or man-induced disasters such as a major industrial accident, large-scale transportation accidents and wars.

4) *Sudden or rapid onset disasters*: these include earthquakes, tsunami, floods, tropical storms, volcanic eruptions and landslides (mostly geological and climatic disasters).

5) *Slow onset disasters*: these include drought, famine, climate change, desertification, deforestation and pest infestation.

Natural disasters	Human-induced disasters
Cyclones	Accidents (road, rail, sea, air)
Cloudburst	Biological disasters, epidemics, health hazards
Drought	Fires (building, coal, mines, oil), urban fires
Earthquakes	Industrial mishaps, gas leak, explosion, sabotage
Floods, flashflood, excessive rains	Nuclear disasters, nuclear weapons, bomb threats, warfare
Landslides	Chemical disasters, oil spills, toxic wastes
Snow avalanches	Electrical/power-related accidents, technical faults, urban heat
Storm surges, tornadoes	Terrorist activities, atom/nuclear bombs, civil strife
Volcanic eruptions	Technological disaster
Wildfires (forest fires)	Dam failure, waterlogging
Tsunamis	Pollution (air, water, soil, noise)
Wind storm, snow storm	Environmental challenges, global warming, sea level rise

Table 1.1. *An overview of natural and human-induced disasters*

1.3. Aggravating factors of disasters

Numerous elements can intensify disasters, leading to increasing their impact and severity in terms of deaths, damage and destruction to the affected population. The following are some typical aggravating factors:

1) *Population growth and rapid urbanization*: if a disaster strikes in a region with a dense population, especially in developing countries, there will be more impact in terms of loss of lives, damage and destruction to structures. Furthermore, if people from rural regions relocate to cities in pursuit of job prospects and safety, there are fewer safe and appealing locations for these migrated people to build their homes. Once more, societal conflicts might result from competing for limited resources.

2) *Poverty and inequality*: poor sections of the population are more vulnerable to disasters compared to wealthy people, as they are able to recover quickly. Economically disadvantaged communities are often forced to live in more vulnerable areas such as flood plains, across borders and other disaster-prone areas; they face greater challenges both in terms of preparing for and recovering from disasters.

3) *Climate change and environmental degradation*: the increasing frequency and intensity of extreme weather events due to climate change can amplify the impact of disasters, leading to more widespread devastation and destruction to people and the environment, which leads to environmental degradation through activities such as deforestation, soil erosion, desertification and poor conservation techniques; these can increase the susceptibility to natural disasters such as floods, landslides and droughts.

4) *Lack of awareness and information*: this is related to inadequate information to people and some people are unaware about safety measures and procedures to be implemented before and after disasters, and may not know about risk reduction measures and steps to take during emergencies.

5) *Inadequate disaster preparedness and infrastructure*: lack of disaster preparedness, early warning systems, evacuation plans and emergency supplies can worsen the impact of disasters, leading to increased vulnerability during disasters. The effects of disasters can be worsened by inadequate infrastructure, such as poorly constructed buildings, roads and bridges, particularly in seismically active zones and cyclone or hurricane-prone locations.

6) *Political instability and conflict*: disasters that strike in the areas impacted by armed conflict or political instability make it more difficult to respond and aggravate the humanitarian situation.

7) *Healthcare system*: inadequacies in healthcare infrastructure and capacity can increase the impact of disasters, particularly in the case of pandemics or during public health emergencies.

1.4. Impact of disasters

Both natural and human-induced disasters have a significant impact on people, economies and the environment, leading to the loss of human lives and causing injuries, often requiring immediate medical attention and emergency response. At times, people may be forced to evacuate their homes due to disasters, leading to temporary or long-term displacement of individuals and families. This can result in homelessness, loss of possessions and challenges to recover again. Physical infrastructure such as buildings, roads, bridges and utilities may be damaged or destroyed during a disaster, hindering normal functioning and requiring extensive reconstruction efforts. The functioning of essential services such as water supply, electricity, communication and transportation can be disrupted. Disasters can result in significant economic losses, including damage to businesses, agriculture and other economic activities. The cost of recovery and rebuilding can be substantial. With regard to natural disasters, such as hurricanes and earthquakes, to man-made disasters, such as oil spills and industrial accidents, the impact can be severe and long lasting, and can cause environmental degradation, such as deforestation, soil erosion, air pollution and contamination of water sources. This can have long-term effects on ecosystems and biodiversity. Other significant impacts of disasters are the disruption of essential services such as healthcare and communications, leading to profound social, health and psychological effects on individuals and communities. Trauma, stress and grief may be experienced due to disruption of social structures. Certain disasters can contribute to the spread of diseases, injuries and mental health issues. Access to healthcare may be compromised, and existing health conditions may worsen. Innovative methods and techniques such as building disaster-resistant buildings and implementing early warning systems have been proven to help effectively mitigate the impact of disasters. By utilizing technologies such as drones and satellite imagery, geoinformatics and remote sensing emergency responders can quickly assess the extent of damage and coordinate rescue efforts more efficiently. Novel research is also being conducted to develop more sustainable and environmentally friendly methods for disaster response, preparation and recovery. For example, scientists are exploring the use of biodegradable materials in temporary shelters and clean-up efforts to minimize the long-term environmental impact of disasters. Additionally, ongoing research used can also improve predictive modeling and risk assessment, allowing communities to better prepare for future disasters to effectively mitigate their impact. By adopting these strategies, communities can become more resilient and better equipped to tackle the challenges of future disasters.

1.5. Hazard meaning and concept

A hazard can be defined as a potential threat or danger that can cause harm to individuals, communities or the environment. It may also be defined as any dangerous physical event, condition, or phenomenon that may cause or has the potential to cause loss of life, injury and damage to property. It can be regarded as an unexpected threat to humans and their property. Also, it is a pre-disaster situation in which some risk of disaster exists, and to be more specific, a hazard is just an event but loss due to this is a disaster. The word "hazard" is derived from old French "*Hasard*" and "*az-har*" in Arabic, which means chance luck. It is a situation where there may be a chance of damage and destruction. Hazard refers to the potential occurrence, in a specific period, and geographic area of a natural phenomenon that may adversely affect human life, property or activity to the extent of causing a disaster. Hazards can arise from a wide range of sources, including natural phenomena, human activities and technological processes. Understanding hazards is crucial for identifying and managing risks to ensure the safety and well-being of individuals and the surrounding environment. To have a clear understanding of hazards, it is essential to grasp the concept and recognize the potential risks they pose. Hazards can be found in various forms and environments, and identifying them is crucial for ensuring safety and minimizing the impact. A hazard becomes a disaster when it affects an area, resulting in loss of life, displacements and homelessness, and damage to infrastructure and property. Understanding hazards involves a comprehensive assessment of their nature, potential impact and the means to control or mitigate their effects. This understanding is achieved through various processes such as hazard identification and risk assessment, taking into account the exposure and vulnerability of individuals or assets.

In conclusion, a hazard represents a potential threat to the well-being of individuals, the environment or property. To gain a better understanding of hazards, it is helpful to explore the concept of hazards with the help of some examples: if a hazard like a cyclone hits an unpopulated area, say an unpopulated coast, it need not be considered a disaster; however; it will be a disaster only if life and property are seriously damaged. Flood is a hazard and is only a disaster when it affects the people and assets. So, we can say that every disaster is a hazard, but it is not necessary that every hazard can be a disaster.

Having a deep understanding of the concept of hazards and being able to identify them is vital for reducing damage and maintaining a safe environment. By recognizing different types of hazards and taking appropriate precautions, individuals can minimize harmful damage and prevent potential risks.

1.6. Types of hazards

There are several different types of hazards, including natural disasters such as earthquakes, floods and hurricanes, as well as man-made hazards such as chemical spills, fires and industrial accidents, biological or behavioral factors that may lead to adverse consequences. Understanding different types of hazards is essential for implementing appropriate safety measures. Each type of hazard presents its own unique set of challenges and risks. These can be grouped into two major categories:

1) *Natural hazards*: natural hazards are caused because of natural phenomena and they are of meteorological, geological and even biological origin, for example, cyclones, tsunamis, earthquakes and volcanic eruptions are exclusively of natural origin.

2) *Man-induced hazards*: man-induced hazards are human caused and are of technological origin, which occur due to human negligence. These are associated with industries or energy generation facilities and include explosions, leakage of toxic waste, pollution, dam failure, wars or civil strife.

1.6.1. *Other types of hazards*

1) *Physical hazards*: these are hazards that arise from the environment and can cause harm to humans or property. Examples include extreme temperatures, sharp objects, noise and radiation.

2) *Chemical hazards*: these hazards involve exposure to chemicals that can be toxic, flammable or corrosive. Examples include cleaning agents, pesticides and industrial chemicals.

3) *Biological hazards*: biological hazards stem from exposure to organisms or substances produced by living organisms. Examples include bacteria, viruses, fungi and allergens.

4) *Behavioral hazards*: these hazards are related to human behavior and can include factors such as negligence, lack of training or disregard for safety protocols, for example, using a mobile phone while driving is a behavioral hazard that increases the risk of accidents.

1.7. Characteristic features of hazard/disaster

Some common characteristics of hazards include the following:

1) *Unpredictability*: hazards often arise unexpectedly or are difficult to predict with certainty; without adequate risk assessment and monitoring, they can appear unexpectedly or build gradually over time, making it difficult to predict when they might occur. Most hazards are difficult to predict and occur suddenly.

2) *Nature*: hazards can be natural, human-induced or a combination of both that may cause damage and destruction.

3) *Location*: the specific physical place or a particular area where a hazard persists and usually occurs in highly vulnerable or disaster-prone areas.

4) *Onset*: hazards/disasters can be sudden, for example, earthquakes, or they can have slow onset, for example, drought.

5) *Intensity*: disasters with great magnitude are more severe and can have more potential to cause damage and disruption.

6) *Frequency*: they can occur at any time; some disasters occur frequently and some are seasonal, causing huge damage in an affected area or a region.

7) *Duration*: this refers to the length of time during which a hazard or dangerous condition exists. The duration of a disaster depends upon the type and effectiveness of a disaster in a particular area.

1.8. Disaster management terminology

– *Capacity*: in disaster management, capacity refers to the ability of individuals, communities, organizations and systems to effectively prepare for, respond to, recover from and mitigate the impact of disasters. It encompasses various elements that contribute to the overall resilience and effectiveness of a community or system in dealing with disasters, involving human resources, skill, education, infrastructure, proper training, etc., to deal with disasters.

– *Capacity development*: capacity development in the context of disasters refers to the process of enhancing the knowledge, skills, resources and capabilities of individuals, communities, organizations and institutions to effectively prepare for, respond to, recover from and mitigate the impacts of disasters. The aim is to strengthen the overall resilience of a community or society to better cope with the challenges posed by natural or human-made disasters.

– *Capacity building*: the process of developing the skills, knowledge, resources and organizational structures required to enhance an individual's or community's ability to effectively address challenges and achieve sustainable development.

– *Coping capacity*: the capacity of individuals, groups and structures to deal with and handle unfavorable circumstances, crises or natural calamities, by utilizing knowledge, available skills and resources to manage adverse conditions of the disasters.

– *Capacity assessment*: evaluation of the existing capacity and resources available to cope with and respond to disasters. This includes assessing the capabilities of emergency services, infrastructure and healthcare systems.

– *Retrofitting*: the process of modifying or upgrading existing structures and infrastructure to make them more resistant to the impacts of disasters, particularly natural hazards like earthquakes, floods, hurricanes or tornadoes.

The goal of retrofitting is to enhance the resilience of buildings and infrastructure, reducing the risk of damage or failure during a disaster event. This is especially important in regions that are prone to specific types of hazards.

– *Resilience*: the capacity of a community, system or society exposed to risks to withstand, resist, adapt and recover from the impacts in a timely and effective way, especially by maintaining and restoring its fundamental structures and functions.

– *Hazard*: a dangerous phenomenon or a condition that may cause loss of life, damage to property, social and economic disruption or environmental damage.

– *Hazard identification*: this is the process of identifying and characterizing the various hazards that could pose a threat to the area. This includes both natural hazards (such as earthquakes, floods and hurricanes) and human-made hazards (such as industrial accidents and chemical spills).

– *Exposure*: exposure refers to the elements (people, houses and infrastructure) and other human assets and property at risk from a natural hazard or human-induced hazard. Exposure is also defined as the state, or situation, of being in contact with something that can develop a potential loss. It may be possible to be exposed but not vulnerable, for example, by living in a floodplain area, but having sufficient means to modify building structure or cope up to mitigate potential loss. To be vulnerable to an extreme event means you are also exposed to a particular hazard.

– *Exposure analysis*: this includes determining the extent to which people, assets and the environment are exposed to the identified hazards. This involves mapping and quantifying the distribution of elements at risk.

– *Vulnerability*: this includes the characteristics and circumstances of a community or a system that makes it susceptible to the damaging effects of a hazard.

– *Vulnerability assessment*: this evaluates the vulnerability of the exposed elements, such as people, buildings, infrastructure and the environment, to the identified hazards. It also assesses the susceptibility and resilience of these elements to potential impacts.

– *Risk*: risk or disaster risk refers to the potential for adverse effects, such as loss of life, damage to infrastructure, economic disruption and environmental degradation, resulting from the interaction between hazards and vulnerable elements. It is a combination of the likelihood of a hazardous event occurring and the potential consequences it may have. It is the probability of quantifiable damage in terms of loss, injury, deaths or any other negative consequences. It means when a community or locality is exposed to a hazard and is likely to be affected by its impact. Risk is the function or product of hazard, exposure and vulnerability.

– *Disaster risk assessment*: this is a systematic process that involves identifying, analyzing and evaluating the potential risks and vulnerabilities associated with natural or human-made hazards. The goal is to understand the potential impact of disasters on a community, region or system and to inform decision-making in disaster risk reduction (DRR) and management.

– *Risk analysis*: this combines information on hazards, vulnerabilities and exposures to assess the overall risk. This often involves the calculation of risk levels, taking into account the probability of a hazard occurring and the potential consequences.

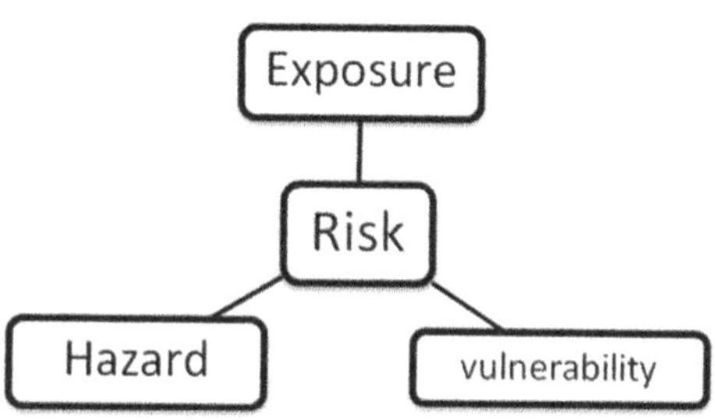

Figure 1.1. *Risk = Hazard × exposure × vulnerability*

1.9. Vulnerability

In disaster management, vulnerability refers to the susceptibility of individuals, communities or systems to the impacts of a disaster. The degree to which a community's infrastructure, services or physical location is likely to be harmed or disturbed by the effects of a certain hazard due to its design, construction or proximity to hazardous terrain or a disaster-prone area can be perceived as their vulnerability. Earthquakes are more likely to affect structures and infrastructure that are vulnerable; therefore, it is important to

adverse disaster events. Disaster management adopts a holistic and multi-sectoral approach, recognizing the interconnectedness of various factors that contribute to disaster risk. The ultimate goal of disaster management is to save lives, reduce suffering and protect livelihoods in the face of adversity. Disaster management aims to build resilient communities and reduce the impact of disasters on individuals and societies.

1.12. Role of different agencies in disaster management

Management of disasters is the responsibility of everyone facing the challenges related to disasters. All the levels of government, whether state, central, NGO, Local people and community play a vital role in the process of management of disaster. It is the responsibility of people, society, communities and local people to manage disasters. By working together, the impact of disasters can be managed effectively to support affected communities in their recovery and rebuilding efforts. Since we cannot completely prevent the upcoming of many disasters, we can reduce their harmful effects by proper management and by being prepared for them before they strike to minimize the loss of life and property. Disasters can be typically managed through a collaborative effort involving various organizations at different levels. The key stakeholders involved in disaster management include:

1) *Government agencies*: national and local government agencies play a crucial role in disaster management, including disaster preparedness, response and recovery. This can involve agencies responsible for emergency management, public safety, healthcare and infrastructure.

2) *Non-governmental organizations (NGOs)*: NGOs often provide vital support in disaster management, offering resources, expertise and ground assistance in the affected areas such as shelter, healthcare and logistics.

3) *International organizations*: organizations such as the United Nations, the World Health Organization and the Red Cross play a vital role in providing global coordination and resources for disaster response and recovery efforts.

4) *Community and civil society groups*: local community organizations and civil society groups play a critical role in disaster management by promoting preparedness, providing support to affected populations and contributing to long-term recovery efforts.

5) *Private sector*: the private sector, including businesses and industry, often contributes resources, logistical support and expertise to disaster management efforts.

1.13. Modern management of disasters

Modern management goes beyond the post-disaster assistance. It now includes both the pre- and post-disaster activities to tackle the particular event of a disaster involving pre-disaster planning, empowering communities through education, awareness, training and participatory approaches to enhance their preparedness, response capabilities, resilience and many other fields. Modern management of disasters involves a holistic and proactive approach that integrates various strategies and technologies to mitigate risks, enhance preparedness and facilitate effective response and recovery. Key components of modern disaster management include:

1) *Risk assessment and mitigation*: this includes utilizing advanced technologies such as geographic information systems (GIS) and remote sensing for comprehensive risk assessment, hazard mapping and identification of vulnerable areas. This information is used to develop and implement targeted mitigation measures.

2) *Early warning systems*: this includes implementing sophisticated early warning systems that leverage real-time data, weather forecasting, seismic monitoring and other technologies to provide timely alerts for impending natural disasters, enabling authorities to take proactive measures and evacuate at-risk populations.

3) *Information and communication technologies (ICT)*: ICT tools help in efficient communication, coordination and data management during disasters, including the use of social media, mobile apps and emergency response platforms to disseminate information, coordinate rescue operations and mobilize resources.

4) *Inclusive and sustainable recovery*: IT includes emphasizing inclusive and sustainable recovery efforts that prioritize the needs of vulnerable groups, promote infrastructure resilience and integrate long-term planning for rebuilding communities more effectively.

1.14. Disaster risk reduction

Disaster risk reduction (DRR) refers to the actions taken to reduce the adverse impacts of natural hazards through systematic efforts to analyze and manage the causes of disaster, including through avoidance of hazards, reduced social and economic vulnerability to hazards and improved preparedness for adverse events. In simple terms, DRR refers to the systematic process of identifying, assessing and reducing the risks of disasters to lessen their adverse impacts on individuals, communities and countries. The goal of DRR is to enhance resilience and minimize the vulnerability of people and their livelihoods to the potential negative consequences of natural and human-made hazards. By implementing comprehensive DRR measures, communities and nations can enhance their ability to anticipate, prepare for, respond to and recover from disasters. DRR can be done through risk assessment by identifying and analyzing potential hazards, vulnerabilities and exposure to determine the level of potential risk to people, infrastructure and the environment. These measures will ultimately reduce the likelihood and overall impact of future disaster events on people. Communities also play a major role in the DRR process through their active participation and empowerment by conducting educational programs to raise awareness about risks and promote preparedness measures. Supporting developing countries and collaborating at the regional and international levels to share knowledge, expertise and resources also helps in building their capacity for DRR. In addition, by designing resilient buildings and structures, promoting insurance schemes and financial mechanisms can also help to withstand or recover quickly from disasters. This will lead to strengthening the capacity of individuals, communities and institutions to effectively manage and respond to both natural and human-induced disasters. DRR considers the probability of harmful consequences or expected losses (deaths, injuries, property, livelihood, economic activity disruption or environmental damage) resulting from interactions between natural or human-induced hazards and vulnerable conditions. Risk can be calculated using the given equation:

$$\text{Risk} = \text{Probability of hazard} \times \text{degree of vulnerability}$$

Management disaster risk involves certain activities:

1) *Prevention*: this includes activities and preventive measures taken to avoid existing and new disaster risks, for example, relocating exposed people and assets away from a hazard-prone region.

2) *Mitigation*: this involves steps to be taken to reduce the harmful impacts of hazards and related disasters, for example, constructing flood defense, planting more trees to stabilize slopes and implementing strict land use policies and construction codes.

3) *Transfer*: The process of formally or informally shifting the financial consequences of a particular hazard from one party to another, whereby a household communities, people or state authority will obtain resources from the other party after a disaster occurs, in exchange for ongoing or compensatory social or financial benefits provided to that party, for example, insurance.

4) *Preparedness*: this includes set of necessary measures taken in advance by people or government. The use of knowledge and capacity of communities and individuals to respond to and recover from the impacts of future hazard or disaster events, for example, disaster preparedness, planning installing early warning systems, identifying evacuation routes and preparing emergency kits and supplies.

1.15. Applications of information technology in DRR

Information technology (IT) has revolutionized communication, bringing within the ambit of connectivity, remote and far-flung and marginalized regions. This helps us in effectively managing disasters.

1) *Decision support and public awareness*: the World Wide Web and the Internet have opened up possibilities of department-specific websites, which provide valuable information in specialized branches of disaster management. There are specialized websites on natural hazards such as earthquakes and cyclones that provide comprehensive information regarding specific natural hazards. Such websites would present all information on the hazard and ways to deal with it. These form important decision support tools (DCS) that facilitate knowledge transfer during critical times.

2) *Information sharing*: the integration of IT with telecommunication interfaces has made possible facilities like video teleconferencing that provide for direct interface between aid givers and official agencies at the emergency site, rendering relief and rescue process in a highly efficient manner.

3) *Policy planning*: IT has greatly aided planning for disaster response and preparedness for DRR. Even generally, policy-making for traffic, transport, forest conservation, urban congestion, etc., is facilitated by spatial imagery through remote sensing.

1.16. Application of GIS in DRR

Another significant development has been the geographical information system (GIS), by which detailed spatial analysis of areas "at risk" is accomplished through satellite imagery. Comprehensive information is collected about the area, which is displayed graphically on a map, highlighting critical facilities and communities at risk, available communication infrastructure, etc., which guides immediate disaster response in the short run, and over the long run, facilitates risk mapping, risk assessment, dissemination of information, public awareness, etc., which aid long-term policy planning for disaster mitigation. The GIS has greatly facilitated response efforts as strategies can be devised based on scientific simulation studies and scenario analysis using information made available through remote sensing. It plays a crucial role in DRR by providing valuable tools and techniques for understanding, analyzing and managing geospatial data related to natural (as well as human-induced) disasters and their potential impact on human populations and infrastructure.

1) *Risk assessment and analysis*: GIS enables the identification and mapping of areas vulnerable to natural disasters such as floods, earthquakes and landslides. It helps in assessing the potential impact of these disasters on communities and infrastructure by analyzing various layers of geospatial data such as elevation, land use and population density.

2) *Early warning systems*: GIS is used to develop early warning systems for natural disasters. By integrating real-time data such as weather patterns, seismic activity and hydrological conditions into GIS platforms, authorities can identify and alert areas at risk, allowing for timely evacuation and disaster response.

3) *Resource management*: GIS assists in the efficient allocation of resources during disaster response and recovery efforts by mapping critical infrastructure, evacuation routes and emergency shelters. This helps in coordinating rescue operations and providing aid to affected areas.

4) *Post-disaster recovery and reconstruction*: GIS also helps in post-disaster assessment and recovery planning by mapping the extent of damage, identifying areas in need of reconstruction and evaluating the effectiveness of recovery interventions.

5) *Policy formulation*: GIS aids in evidence-based decision-making for DRR policies and land use planning by providing spatial analysis and visualization of vulnerability and exposure to natural hazards.

1.17. Disaster management cycle

The disaster management cycle, also known as the disaster management continuum, is a conceptual framework that outlines the various phases of dealing with disasters, from preparedness to recovery. This cycle helps emergency management professionals and organizations to plan, coordinate and implement strategies for disaster mitigation and response. The typical phases of the disaster management cycle include the following:

1) *Preparedness*: this phase includes planning, training, drills, public awareness campaigns, infrastructure development and resource allocation. Preparedness enhances the capacity of communities and organizations to respond effectively to disasters.

2) *Mitigation*: this includes implementing measures to reduce or eliminate the impact of hazards, such as building codes, land-use planning and infrastructure improvements to minimize the impact of disasters on people and property.

3) *Response*: this includes immediate actions that are taken to save lives and prevent further damage during and after a particular disaster. It also includes search and rescue, medical care, evacuation and emergency shelter to protect lives and property, meet basic human needs and stabilize the situation.

4) *Recovery*: long-term efforts to restore the affected community to normal or improved conditions. This involves rebuilding infrastructure, providing financial assistance and addressing social and economic issues to restore the community to a state of normalcy or improved resilience.

Figure 1.2. *Disaster management cycle*

1.18. Disaster Management Act

The Disaster Management Act, 2005 was enacted on December 23, 2005 and extends to the whole of India. It received the assent of the President of India on the same date. It serves as a crucial legal framework for managing disasters and providing support to those affected in India, and consists of all the definitions of disaster and Disaster Management concepts. The Disaster Management Act of 2005 has provided the legal and institutional framework for Disaster Management in India at national, state and district levels that focus on providing effective management of disasters and related matters. The main objective of this act is to build a safe and disaster-resilient India through prevention/mitigation, preparedness and response to provide technical and financial activities support and other commercial activities. The Act consists of 11 chapters and 79 sections, outlining various measures for disaster management and related issues. Its main focus is to assist people affected by disasters, helping them regain their lives to prevent, mitigate and prepare for future disasters.

1.18.1. *Chapters of Disaster Management Act, 2005*

The chapters of Disaster Management Act, 2005, include the following:

1) Preliminary chapter.

2) The National Disaster Management Authority (NDMA).

3) State Disaster Management Authority (SDMA).

4) District Disaster Management Authority (DDMA).

5) Measures by the Government for Disaster Management.

6) Local authorities.

7) National Institute of Disaster Management (NIDM).

8) National Disaster Response Force (NDRF).

9) Finance.

10) Offences and penalties.

11) Miscellaneous.

1.18.2. *Disaster Management Act, 2005 framework*

The National Disaster Management Act of 2005 in India establishes a framework for disaster management at the national, state and district levels. The act outlines the roles and responsibilities of various bodies at different levels involved in disaster management. Here are some of the key bodies mentioned in the National Disaster Management Act, 2005:

1) *National Disaster Management Authority (NDMA)*: NDMA is the apex body responsible for formulating policies, plans and guidelines for disaster management in India. It is headed by the Prime Minister of India. The NDMA coordinates disaster response and mitigation efforts at the national level. The NDMA is also mandated to recommend guidelines for minimum standards of relief to be provided to persons affected by the disaster for proper allocation of funds.

2) *State Disaster Management Authority (SDMA)*: each state in India has its own State Disaster Management Authority, headed by the chief minister. In the case of union territory with no assembly, for example, Delhi and Puducherry, the LG or Administrator is the chairman of SDMA. The

state-level implementation of the plans and policies established by the NDMA is the responsibility of the SDMA. It creates the State Disaster Management Plan and organizes disaster response efforts.

3) *District Disaster Management Authority (DDMA)*: at the district level, the DDMA is responsible for implementing disaster management plans and coordinating response activities. It is headed by the District Magistrate (Deputy Commissioner) or Collector.

4) *National Executive Committee (NEC)*: the NEC is responsible for coordinating the implementation of policies and plans for disaster management. It is chaired by the Union Home Secretary.

5) *State Executive Committee (SEC)*: similar to the NEC, the SEC at the state level coordinates the implementation of disaster management plans within the state. It is chaired by the Chief Secretary of the state.

6) *Local authorities*: municipalities, panchayats and other local bodies are involved in disaster management at the grassroots level. They work under the guidance of the DDMA and play a significant role in coordinating local-level response efforts.

1.19. Disaster preparedness plan

Disasters can disrupt the normal functioning of a community, resulting in significant damage to property and lives, causing a breakdown in social and economic systems. Individuals and authorities are often caught off guard, lacking the initiative and direction needed to respond effectively to such crises. This can lead to delays in relief efforts, further exacerbating the situation. However, having a well-developed and tested disaster preparedness plan in place can significantly mitigate the impact of disasters and enhance disaster resilience through effective preparedness planning. A comprehensive disaster preparedness plan equips officials with a set of clear instructions to follow and enables them to issue directives to their teams and the affected population. This streamlined approach not only expedites rescue and relief operations but also boosts morale among those affected. Moreover, such plans are invaluable in pre-disaster situations, guiding officials when warnings are issued and eliminating the need for time-consuming consultations with senior officers for approval. The importance of having a disaster preparedness plan is evident from national and international experiences, where inadequate planning or the absence of a plan has led to more severe consequences from disasters. A well-structured plan includes measures to be implemented before, during and after a disaster strikes. It outlines the inventory of available resources, assigns responsibilities, establishes coordination mechanisms among government bodies and provides guidelines for collaboration with NGOs, social workers and international organizations. A well-designed disaster preparedness plan not only enhances the resilience of communities but also ensures a coordinated and efficient response to disasters, minimizing loss of life and property.

By creating awareness and empowering communities to anticipate and respond to disasters, a preparedness plan aims to minimize the impact of emergencies. Long-term measures such as constructing river embankments, retrofitting buildings, establishing communication systems, setting up relief centers and implementing land use regulations are also included in the plan. Regular review and revisions ensure the plan's practicality and effectiveness based on lessons learned from past experiences.

Key characteristics of an effective preparedness plan are as follows:

1) *Clarity of aim*: the plan should have a clearly defined objective that aligns with the overall goal of disaster preparedness.

2) *Realism*: it should be based on a realistic assessment of potential disaster threats, community vulnerability and available resources to counter such threats.

3) *Level of the plan*: the plan should be tailored to the specific level of operation, whether at the local village, district or regional level.

4) *Flexibility*: given the unpredictable nature of disasters, the plan should be adaptable and able to respond to varying scenarios swiftly.

1.20. Questions

1) Define the following terms:

a) Disaster and risk

b) Resilience and exposure

c) Capacity and capacity development

2) Define Vulnerability; also discuss its types.

3) Distinguish between hazard and disaster.

4) Explain in detail the characteristic features of hazards.

5) What are the two major types of disasters? Name three examples in each category.

6) Write in detail about disaster risk reduction (DRR).

7) Describe the disaster management cycle and its all phases.

8) Write a detailed note on the Disaster Management Act.

9) Explain the concept of risk assessment in disaster management and discuss its importance in proactive planning.

10) Describe the principles of disaster risk reduction and explain how they contribute to sustainable development.

11) Explain the role of technology in disaster management, highlighting both its benefits and limitations.

12) Discuss the applications of GIS and remote sensing in disaster risk reduction.

13) What are the aggravating factors of disasters?

14) Discuss the importance of preparedness in disaster management.

15) Describe the concept of disaster management and its scope.

16) What do you understand by a disaster preparedness plan?

17) What is the importance and significance of a disaster preparedness plan?

18) What are the main characteristics of a disaster preparedness plan?

19) Discuss the role of government agencies at the local, national and international levels in disaster management.

20) Describe the role of non-governmental organizations (NGOs) and volunteer groups in disaster management.

1.21. Multiple-choice questions (MCQs)

1) What is the primary goal of disaster management?

a) Recovery

b) Preparedness

c) **Mitigation**

d) Response

2) What is the main purpose of a disaster response plan?

a) Reduce the risk of disasters

b) Minimize the impact of disasters

c) **Coordinate actions during and after a disaster**

d) Implement long-term recovery efforts

3) The process of identifying potential hazards and assessing their impacts is known as:

a) Response planning

b) **Risk assessment**

c) Emergency preparedness

d) Disaster recovery

4) Which organization plays a crucial role in coordinating disaster response efforts at the national level in many countries?

a) **United Nations**

b) World Health Organization

c) Red Cross

d) Federal Emergency Management Agency (FEMA)

5) What does the term “Shelter in Place” mean in disaster management?

a) Evacuating to a designated shelter

b) Taking cover in a safe location

c) **Staying indoors and taking refuge**

d) Moving to higher ground

6) The "Incident Command System" (ICS) is a standardized approach for:

a) Disaster recovery

b) Communication during disasters

c) **Incident management and coordination**

d) Risk assessment

7) What is the purpose of conducting drills and exercises in disaster management?

a) Assessing the damage after a disaster

b) **Testing the effectiveness of response plans**

c) Providing immediate relief to affected populations

d) Identifying potential hazards

8) The "Golden Hour" in disaster management refers to:

a) The first hour after a disaster occurs

b) **The critical period for decision-making and action**

c) The time when recovery efforts begin

d) The hour of maximum impact during a disaster

9) What is the purpose of a Community Emergency Response Team (CERT)?

a) **Search and rescue operations**

b) Conducting risk assessments

c) Managing recovery efforts

d) Providing medical care during disasters

10) Which disaster management phase involves restoring the community to normal or improved conditions?

a) Preparedness

b) Response

c) **Recovery**

d) Mitigation

11) What is the primary role of the National Disaster Response Force (NDRF)?

a) Providing financial assistance to disaster victims

b) Researching disaster management

c) **Undertaking specialized response and rescue operations**

d) Coordinating international disaster relief efforts

12) The "Triage" process in disaster management is related to:

a) **Sorting and prioritizing casualties for medical treatment**

b) Assessing the structural damage after a disaster

c) Evacuating people from affected areas

d) Distributing relief supplies

13) What is the purpose of a Hazard Vulnerability Assessment (HVA)?

a) **Assessing the risk of hazards to a community**

b) Providing immediate relief to disaster victims

c) Coordinating international disaster response efforts

d) Evaluating the effectiveness of recovery plans

14) The "Chain of Command" in disaster management refers to:

a) A series of interconnected emergency shelters

b) **The hierarchy of authority and communication within response teams**

c) The distribution of relief supplies to affected areas

d) A network of international disaster relief organizations

15) The term "disaster risk reduction" (DRR) is closely associated with:

a) **Minimizing the impact of disasters on vulnerable populations**

b) Providing immediate relief to disaster victims

c) Conducting post-disaster damage assessments

d) Coordinating international disaster response efforts

16) What is a "safe zone" in disaster management?

a) A designated area for emergency drills

b) A location for storing relief supplies

c) **A secure area for affected populations during a disaster**

d) A zone with reduced risk of natural disasters

17) The term "Crisis Communication" in disaster management refers to:

a) Communicating during the recovery phase

b) Providing immediate relief to disaster victims

c) **Communicating information during and after a disaster**

d) Coordinating international disaster response efforts

18) The "Sendai Framework for disaster risk reduction" was adopted by:

a) United Nations Security Council

b) **United Nations General Assembly**

c) World Health Organization (WHO)

d) International Red Cross

19) Which phase of disaster management involves educating the public and raising awareness about potential risks?

a) **Preparedness**

b) Response

c) Mitigation

d) Recovery

20) The term "evacuation route" in disaster management refers to:

a) A designated path for emergency drills

b) The route taken by relief supplies to affected areas

c) **The path used by people to leave an area during a disaster**

d) A route for transporting injured individuals to medical facilities

21) What is the purpose of the "Hyogo Framework for Action"?

a) Addressing climate change issues

b) **Enhancing disaster risk reduction efforts**

c) Coordinating international humanitarian aid

d) Conducting research on disaster management

22) The term "FEMA" stands for:

a) **Federal Emergency Management Agency**

b) Foreign Emergency Medical Assistance

c) Fund for Emergency Humanitarian Aid

d) Federation of Emergency Management Associations

23) Widespread loss of life and property due to natural or manmade causes is termed a:

a) **Disaster**

b) Hazard

c) Extreme loss

d) Vulnerability

24) The Disaster Management Act was made in:

a) 2006

b) 2003

c) **2005**

d) 2009

25) The level of risk of a disaster depends on:

a) Nature of the hazard

b) Vulnerability of the elements which are affected

c) The economic value of the element which is affected

d) **All of the above**

26) Any natural process with the potential to cause harm is termed a natural hazard.

a) **True**

b) False

27) A physical, social, economic and environmental condition that make a society susceptible to the adverse effects of hazard is known as:

a) Resilience

b) Capacity

c) **Vulnerability**

d) Mitigation

28) Disaster risk is a function of:

a) Hazard

b) Vulnerability

c) Exposure

d) **All of above**

29) The word disaster comes from the:

a) Greek word

b) Latin word

c) **French word**

d) German word

30) Where is NIDM located?

a) Kanpur

b) Bengaluru

c) **New Delhi**

d) Mumbai

31) J&K is exposed to various hazards mainly due to what?

a) Location

b) Geomorphology

c) Climate

d) **All of above**

32) Who should be the first responders in a disaster management situation?

a) Central government

b) **Community**

c) Local government

d) State government

33) Disaster management includes:

a) Mitigation

b) Reconstruction

c) Rehabilitation

d) **All of above**

34) Which of the following groups of people are more vulnerable in the event of disaster?

a) Men, women, boys

b) Men, boys, old people

c) **Women, children, old people**

d) None of the above

35) What are the consequences of disaster on a society?

a) Loss of life

b) Damage to property

c) Environmental degradation

d) **All of above**

36) When do we celebrate International Day for Natural Disaster Reduction?

a) March 1

b) April 15

c) September 12

d) **October 13**

37) What are the major hazard /disaster management-related activities?

a) Preparedness

b) Response

c) Recovery

d) **All of the above**

38) What is a disaster an occurrence of?

a) Disrupting the normal conditions of existence

b) Causing a level of suffering

c) Damage to infrastructure

d) **All of the above**

39) What are hazards and disasters mainly classified as?

a) Physical and chemical

b) **Natural and human-induced**

c) Physical and human

d) Social and cultural

40) A hazard is a situation where there is:

a) The threat of natural calamity

b) **The threat to property and lives from calamities**

c) The threat for consequences of the disaster

d) All of the above

41) Disaster is an event arising out of:

a) **Result of hazard event**

b) Causes of hazard event

c) Causes of disaster event

d) All of the above

42) Typical examples of man-made disasters are:

a) Chemical explosion

b) Leakage of toxic waste

c) War and civil strife

d) **All of the above**

43) The role of which agency is important in disaster prevention?

a) **Media**

b) Police

c) Government officials

d) Public

44) What are the important measures to be taken at the community level of disaster preparedness?

a) Increased awareness

b) Provision of early and timely warning

c) Land use planning

d) **All of the above**

45) Which of the following agencies are the national disaster risk reduction stakeholders?

a) Central government

b) National disaster offices

c) Private sector

d) **All of the above**

46) Structural and nonstructural measures are the elements of which category of risk reduction measures?

a) Socioeconomic measures

b) **Physical measures**

c) Environmental measures

d) Post-disaster measures

47) What are the major types oriented to mitigation measures of hazards and disasters?

a) Public mitigation measures

b) Community mitigation measures

c) **Both a and b**

d) None of the above

48) What are the main prescribed procedures for standard operating?

a) Search and rescue

b) Medical assistant

c) Casualty management

d) **All of the above**

49) Which phase of disaster management involves measures taken to minimize the impact of a disaster?

a) Preparedness

b) Response

c) Recovery

d) **Mitigation**

50) The steps needed to be taken for any risk of natural or man-made disaster are:

a) Proper building construction

b) Better planning

c) Awareness among the stakeholders

d) **All of the above**

51) Disaster management is aimed at:

a) Resettling people in the closest unaffected urban area

b) Collection of valuable data for future management objectives

c) Strengthening sewage and drinking water treatment facilities to resist the impact of a future disaster

d) **Restoring a community's services, facilities and residences to pre-disaster levels**

52) What are the roles of NGOs in the disaster-related preparedness stage?

a) Public awareness and education

b) Vulnerability and risk assessment

c) Forming and training DNTs, introducing alternative technologies

d) **All of the above**

53) Mock drills and exercise to create awareness among people are conducted at:

a) State level

b) District level

c) GP/Village level

d) **All of the above**

54) Disaster management covers:

a) Maintaining control over disasters

b) Reducing the effects of disasters

c) Briefing of government officials about disasters

d) **All the above**

55) The primary goal of a disaster preparedness plan is:

a) **To protect the population**

b) To protect valuable resources

c) To keep communication lines open

d) To protect environmental health personnel

56) For coordinating disaster management activities for all natural hazards, the nodal agency at the central government is the:

a) **Ministry of Home Affairs**

b) Ministry of Rural Development

c) Ministry of Communication

d) Ministry of Urban Ministry

57) What are the three phases of disaster management planning?

a) Evacuating, rebuilding and re-branding

b) Preparation, planning and perception

c) Planning, evacuating and recovery

d) **Preparation, response and recovery**

58) Full form of DDMA is:

a) **District Disaster Management Authority**

b) District Disease Management Authority

c) District Dam Management Authority

d) District Duck Management Authority

59) Which of the following best defines "risk" in the context of a system or environment?

a) **The probability of a specific hazard or threat occurring**

b) The potential consequences or impacts of an event

c) The level of exposure to potential harm or loss

d) The ability of a system to adapt and recover from disturbances

60) What is the full form of NEC?

a) National Eternal Committee

b) **National Executive Committee**

c) National Election Commission

d) National Engineering Course

2

Understanding Natural Disasters

Natural disasters refer to the catastrophic events resulting from natural processes on Earth that are frequently abrupt and powerful, causing significant damage, casualties and fatalities, while also disturbing the balance of daily life. People are thus forced to reside in vulnerable locations due to several socioeconomic causes, including population growth. India's unique geo-climatic position makes the country particularly vulnerable to natural disasters. India is a vast peninsula of sub-continental size, surrounded by water on three sides and has the Himalayan range on the fourth side, which has some of the tallest mountains in the world. This is why India has to face a very large number of disastrous events of geological, oceanic or climatic origin. The affected community's suffering is typically exacerbated by poverty, dense population, inadequate infrastructure, close proximity to rivers, mountains, or the sea, a lack of community cohesion and ineffective government. Approximately 60% of all disasters are caused by flooding and strong winds. The landmass of the subcontinent is susceptible to earthquakes to the extent of around 54%, and the country is susceptible to recurrent flooding to the extent of approximately 4 crore ha. A natural disaster is a significant disruption to normal human existence and the economic system that may arise from an abrupt, catastrophic incident or a gradual process. These catastrophic events can cause significant disruption to the environment, people and their property. These are often unpredictable and can occur suddenly, posing a threat to communities and ecosystems. The severity of a disaster is assessed based on the disruption to the normal pattern of life and impacts such as loss of life, loss of property, injury, hardship and adverse effects on health. At this stage, the community needs shelter, food, clothing, medical assistance and social care. Some common types of natural disasters include earthquakes, landslides, floods, volcanic eruptions, tsunamis, cyclones, storms (wind, sea and storm surges), tornadoes, droughts and wildfires. Natural disasters can have devastating effects on communities, causing loss of life, displacement of populations, destruction of infrastructure and long-term environmental impacts. Preparedness, early warning systems and effective response measures are crucial in minimizing the impact of these events. Therefore, understanding these natural disasters and building a disaster-resilient community is crucial for reducing the damage caused by these events. Natural disasters encompass a wide range of events, including earthquakes, hurricanes, floods, tsunamis, wildfires, etc. Each type of disaster has unique characteristics, impacts and mitigation measures. By understanding the nature of different natural disasters, their potential effects and implementing effective mitigation strategies, communities and governments can work toward reducing the impact of these events and building resilience for future occurrences.

2.1. Earthquakes

2.1.1. *Meaning and concept*

Earthquakes are sudden geological events below the Earth's surface that result in the generation of waves that travel far and wide, creating vertical and horizontal vibrations. The severity of the impact may vary depending upon the magnitude and intensity of an earthquake, which in turn depends on the amount of energy released at a point where the seismic event took place below the surface of the Earth. Earthquakes occur suddenly, and thus there is no dependable technique for the prediction of earthquakes as of yet. Earthquakes are natural and one of the worst disasters caused due to the displacement of rocks and a sudden release of energy in the Earth's crust that generates seismic waves to propagate in all directions. This release of energy along a fault is typically the result of tectonic forces, which are generated by the movement of the Earth's lithospheric plates. In simple terms, earthquakes on the surface are manifested by the shaking of the ground and are one of the most dangerous natural hazards causing widespread destruction and loss of human lives. In addition, they can trigger secondary disasters, such as tsunamis, floods and volcanic eruptions. Earthquakes are unpredictable and are capable of causing large-scale damage and killing thousands of people. The Indian subcontinent's tectonic setting and the interactions of major geological plates are the main causes of frequent earthquakes and seismic activities. The vulnerability to various earthquake hazards requires long-term planning to reduce its impact. An earthquake is a rapid, intense shaking of the ground as a result of movements in the Earth's crust. These shakings usually cause tremendous destruction. India's geographic statistics indicate that approximately 58% of the country is susceptible to earthquakes. Out of the total earthquake-prone areas, 12% are prone to very severe earthquakes, 18% to severe earthquakes and 25% are prone to damageable earthquakes.

Figure 2.1. *An earthquake. For a color version of this figure, see www.iste.co.uk/akbar/disastermanagement.zip*

2.1.2. *Earthquake-related terms*

– *Focus (or hypocenter)*: the actual location within the Earth where the earthquake originates, or the place of origin of an earthquake. This is where the rocks break and slip, releasing energy.

– *Epicenter*: the point on the Earth's surface directly or vertically above the focus is called the epicenter of an earthquake. Maximum damage is observed near the epicenter.

– *Seismology*: the scientific study of earthquakes and seismic waves is known as seismology. Examining the propagation of elastic waves through the Earth, especially waves produced by natural phenomena like earthquakes or volcanic activity, aims to comprehend the behavior of the Earth's interior.

– *Isoseismic line*: a line on a map that connects points experiencing the same level of seismic intensity during an earthquake.

– *Seismograph*: a seismograph, also known as a seismometer, is a scientific instrument designed to detect and record vibrations or seismic waves reaching the Earth's surface generated by movements within the Earth. It helps scientists to study and monitor seismic events.

– *Seismogram*: a seismogram is a graphical representation of ground motion/movement recorded by a seismograph or by a seismometer during an earthquake.

– *Seismic waves*: vibrations that travel outward from the earthquakes focus and shake the ground. Seismic waves are of three types: primary waves (P-waves), secondary waves (S-waves) and surface waves.

– *P-waves*: these are also known as primary or compressional waves that travel fast through the Earth and are the first to arrive at the surface. They can pass through solid, liquid and gaseous materials. These waves are similar to sound waves and cause rock particles to move in the same direction as the wave.

– *S-waves*: these are also known as secondary or shear waves and follow P-waves. These waves travel only through solid materials. They move rock particles perpendicular to the direction of the wave and are slower than P-waves.

– *Surface waves or L-waves*: these are also known as long period waves. These types of waves travel along the Earth's surface and are the last waves to report on seismograph that cause most damage and destruction during an earthquake.

– *Magnitude*: the magnitude of an earthquake is a measure of its size or the total amount of energy released during a particular earthquake, and it is usually reported on the Richter scale ranging from 0 to 10.

– *Intensity*: it is the degree of damage and destruction caused by the particular earthquake at specific locations and is measured on the modified Mercalli intensity (MMI) scale ranging from 1 to 12.

– *Ring of Fire*: the Pacific Ring of Fire encompasses the coasts of several countries bordering the Pacific Ocean. It is home to approximately 75% of the world's active and dormant volcanoes, as well as about 90% of the world's earthquakes occur along the Ring of Fire. The Ring of Fire is sometimes called the circum-Pacific belt.

2.1.3. *Types of earthquakes*

Earthquakes can be classified into different types based on various factors such as their origin, depth of focus and the geological features involved. The different types of earthquakes are as follows.

2.1.3.1. *Classification based on causal factors*

– *Tectonic earthquakes*: these types of earthquakes occur due to the sliding of rocks along a fault plane or due to the movement of tectonic plates.

– *Volcanic earthquakes*: these types of earthquakes are associated with volcanic activity and occur due to the movement of magma beneath the Earth's surface and are more common in areas with active volcanoes.

– *Collapse earthquakes*: these earthquakes are caused due to intense mining activity by the collapse of underground caverns or mines causing minor tremors.

– *Explosion earthquakes*: these types of earthquakes are induced by the explosion of nuclear or chemical devices.

– *Human-induced earthquakes*: these earthquakes are a result of human activities such as mining, reservoir-induced seismicity (due to the filling of large reservoirs behind dams) and hydraulic fracturing (fracking).

2.1.3.2. *Classification based on depth of focus*

– *Moderate or shallow focus earthquakes*: this includes earthquakes having depth of focus from the ground in between 0 and 70 km.

– *Intermediate focus earthquakes*: the depth of focus lies in between 70 and 300 km.

– *Deep focus earthquakes*: earthquakes having a depth of focus ranging between 300 and 700 km.

2.1.4. *Causes of earthquakes*

Earthquakes are caused when there is a sudden release of energy in the Earth's crust that creates seismic waves. The movement of tectonic plates beneath the surface of the Earth is the main cause of earthquakes. The lithosphere, or outer layer of the Earth, is separated into tectonic plates, which are huge and small fragments. The primary causes of earthquakes are as follows:

1) *Tectonic plate movements*: relative movement of plates is the main cause behind earthquakes. These plates are constantly in motion due to the heat-driven convection currents in the mantle. These plates can slide past one another, move apart, or collide when they come into contact with one another.

2) *Subduction zones*: majority of earthquakes occur where one tectonic plate can be forced beneath another in a process called subduction. As a plate is subducted into the Earth's mantle, intense pressure and heat can cause it to melt and generate magma. This magma can lead to the formation of volcanoes and can also trigger earthquakes.

3) *Volcanic activity*: earthquakes are commonly associated with volcanic activity. As magma rises beneath the Earth's surface, it can create pressure and fractures to push the crustal surface from below with force leading to earthquakes.

4) *Folding and faulting*: the horizontal and vertical movements caused by endogenic forces result in the formation of faults and folds, which in turn cause dislocation of rock blocks and ultimately cause earthquakes of varying magnitude and intensity.

5) *Human-induced activities*: certain human activities, such as underground mining, blasting of rocks, nuclear explosions, filling of large reservoirs behind dams, and hydraulic pressure (fracturing) oil and gas extraction, can induce seismic activity and tremors. While these earthquakes are generally smaller than natural ones, they can still cause damage.

6) *Isostatic rebound*: earthquakes can result from the Earth's crust adjusting to changes in the distribution of mass. For example, the melting of glaciers and the unloading of ice can lead to isostatic rebound, causing stress adjustments in the Earth's crust and triggering earthquakes.

7) Other natural phenomena, such as the collapse of underground caverns, landslides and the impact of meteorites, can also generate seismic waves and produce localized earthquakes.

2.1.5. *Types of plate boundaries*

Earthquakes are primarily associated with the movement of tectonic plates at plate boundaries. There are several types of plate boundaries, and the interactions at these boundaries contribute to different types of earthquakes. Different types of plate boundaries are described below:

– *Divergent boundaries*: these are the boundaries where tectonic plates move away from each other. The sites or locations where the plates diverge from each other are called spreading sites. This movement creates a gap that allows magma to rise from the mantle, solidify and form a new crust. As the plates separate, earthquakes can occur due to the movement of rocks leading to shallow-focus earthquakes.

– *Convergent boundaries*: these are the boundaries where two tectonic plates move toward each other. When an oceanic plate collides with a continental plate or another oceanic plate, the denser oceanic plate may subduct beneath the other plate. The place where the subduction of plates takes place is known as subduction zone or Benioff zone and is associated with powerful earthquakes.

– *Transform fault boundaries*: these are the boundaries where the two tectonic plates slide past each other horizontally. Due to sliding, the crust is neither produced nor destroyed. The friction between these plates prevents smooth sliding, causing stress to build up.

2.1.6. *Earthquake risk zones of India*

A large portion of India is vulnerable to earthquake activity of varying magnitude. Most of the highly vulnerable areas are in the Himalayan and sub-Himalayan regions, Andaman and Nicobar Islands and the Kutch area of Gujarat. India is seismically active due to its location near the boundary of the Indian Plate and the Eurasian Plate. The collision and interaction of these tectonic plates have led to the formation of several fault lines, making different regions of India prone to earthquakes. The seismic risk zones in India are categorized based on the seismic zone map provided by the Bureau of Indian Standards (BIS). The BIS has classified the country into four seismic zones (by unifying the first two zones) ranging from Zone II to Zone V, with Zone V being the most seismically active. Here is an overview of the seismic zones in India:

1) **Seismic Zone II:** this zone represents areas of low seismic risk. Some parts of northern and eastern India fall into this category. However, it is important to note that even in Zone II, buildings and infrastructure need to adhere to seismic design codes to ensure safety.

2) **Seismic Zone III:** this zone includes regions of moderate seismic risk. Many major cities, including Delhi, Mumbai, Kolkata and Chennai, fall into Zone III. Building codes in these areas incorporate seismic design considerations to mitigate the impact of earthquakes.

3) **Seismic Zone IV:** Zone IV is characterized by high seismic risk. Cities such as Ahmedabad, Srinagar, Jammu and parts of Himachal Pradesh and Uttarakhand are in this zone. The building codes in Zone IV are more stringent to ensure structures can withstand stronger seismic forces.

4) **Seismic Zone V:** this zone represents areas of very high seismic risk. The Himalayan region, including parts of Jammu and Kashmir, Himachal Pradesh, Uttarakhand and north-eastern states, falls into Zone V. These areas are most susceptible to severe earthquakes.

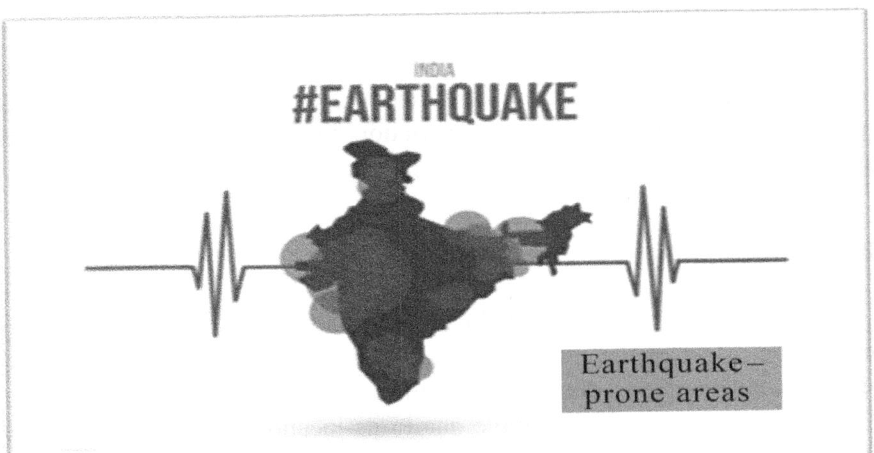

Figure 2.2. *Areas vulnerable to earthquakes in India. For a color version of this figure, see www.iste.co.uk/akbar/disastermanagement.zip*

2.1.7. *Some damaging earthquakes in India*

India has a very long history of earthquakes. The occurrence of earthquakes is not a new phenomenon for most of the northern parts of the country. In the recent past, India experienced major earthquakes at Uttarkashi (1991), Latur (1993), Jabalpur (1997), Bhuj (2001) and Kashmir (2005). In this section, we will discuss in detail a few of these recent earthquakes.

1) *Uttarkashi earthquake (1991)*: the Uttarkashi earthquake occurred on October 20, 1991, in the Uttarkashi district of the Indian state of Uttarakhand (formerly part of Uttar Pradesh). The earthquake had a magnitude of 6.8 and resulted in significant damage to buildings, infrastructure and the loss of lives. The epicenter of the earthquake was located near the town of Uttarkashi. Several towns and villages in the region were severely affected, including Uttarkashi, Bhaironghati and Dharasu. The earthquake caused widespread destruction, leading to the collapse of buildings, landslides and disruption of communication and transportation systems. The exact number of casualties and the extent of the damage varied across different reports, but it is estimated that hundreds of people lost their lives and many more were injured.

2) *Latur earthquake (1993)*: an earthquake of moderate magnitude (6.4 on the Richter scale) struck the Marathwada region of Maharashtra state on September 30, 1993. The impact of this earthquake was felt in the adjoining states of Andhra Pradesh and Karnataka. However, the severely affected areas were mainly the Latur and Osmanabad districts of Maharashtra.

3) *Bhuj earthquake (2001)*: on January 26, 2001, when the nation had just started the Republic Day celebrations, a devastating earthquake struck near Bhuj in the Kachchh region of Gujarat in the morning. The magnitude of this earthquake was 6.9 on the Richter scale and it was the borderline earthquake between the "moderate" and "great" categories. Hitting a prosperous region of the country, its impacts were truly disastrous.

4) *Kashmir earthquake (2005)*: the Kashmir earthquake struck on October 8, 2005, with a magnitude of 7.6 on the Richter scale and was one of the deadliest earthquakes in the history of the region. Pakistan-administered Kashmir's Muzaffarabad town was close to the epicenter. The earthquake caused widespread devastation, leading to the collapse of buildings, bridges and infrastructure. It also triggered landslides in mountainous regions, further exacerbating the impact. The earthquake

affected not only Kashmir but also parts of north-western India and Eastern Afghanistan. The human toll was significant, with estimates of casualties ranging from 70,000 to 80,000 people. Many more were injured, and countless homes and structures were destroyed.

India is highly vulnerable to earthquakes and has been the victim of very severe earthquakes, which are presented in Table 2.1.

Date	**Location**	**Magnitude**
June 16, 1819	Kutch, Gujarat	8.0
January 10, 1869	Cacher, Assam	7.5
May 30, 1885	Sopore, J&K	7.0
June 12, 1897	Shillong Plateau	8.7
April 4, 1905	Kangra, Himachal Pradesh	8.0
July 8, 1918	Srimangal, Assam	7.6
October 20, 1991	Uttarkashi, U.P. Hills	6.6
September 30, 1993	Latur, Maharashtra	6.3
March 29, 1999	Chamoli	6.8
January 26, 2001	Bhuj, Gujarat	6.9
October 8, 2005	Muzaffarabad, J&K	7.4

Table 2.1. *Significant earthquakes in India (source: http://www.ind.ernet.in/section/seismo/static/signif.htm)*

2.1.8. *Effects of earthquakes*

Earthquakes occur suddenly without any warning. It is not possible to forecast earthquakes in terms of exact time of occurrence, place and magnitude of an earthquake. These events have more effect on dense populations; it is not earthquakes that kill the people, but the falling buildings that do. The nature and impact of an earthquake disaster is dependent on several factors including its magnitude, geological and soil conditions, depth distance from epicenter and location of the fault. Earthquakes can have various effects causing casualties, injuries or even can lead to death of people. The impacts of earthquakes can be wide-ranging and devastating. Here are some of the hazardous effects of earthquakes:

1) *Structural damage*: earthquakes can cause severe damage to buildings, bridges, roads and other infrastructure. High-magnitude earthquakes can lead

to complete collapse and destruction, posing danger to humans and can result in loss of life, injuries and displacement of people.

2) *Ground rupture*: the movement of the Earth's crust during an earthquake can cause visible ruptures in the ground, changes in underground water level, dam breach and displacement of land leading to widespread damage to roads, pipelines and other structures. In addition, earthquakes also cause changes in the course of rivers or streams.

3) *Economic and environmental impact*: the destruction of infrastructure, homes and businesses can have significant economic repercussions, impacting local and national economies. In addition, earthquakes can lead to environmental damage, including contamination from ruptured industrial facilities and hazardous material spills.

4) *Psychological impact*: the trauma and fear associated with experiencing an earthquake or its aftermath can have long-lasting psychological effects on individuals and communities.

5) *Disruption of services*: earthquakes are dangerous and may lead to disruption of essential services such as water, electricity gas lines transport and communication networks.

6) *Secondary hazards*: when an earthquake strikes, it can lead to secondary hazards such as fires due to rapturing of gas pipes and electrical shorts. Broken gas lines can cause leakages of gas and result in explosions and fires, posing additional risks to the affected areas. It can also lead to other natural and manmade disasters such as floods, tsunamis, landslides and soil liquefaction, resulting in damage and destruction to the people and their property.

7) *Tsunamis*: underwater earthquakes through seawater generate high seismic waves and have the potential to trigger tsunamis, which are large ocean waves capable of causing widespread coastal destruction and flooding.

8) *Landslides and avalanches*: earthquakes can trigger landslides and avalanches, particularly in hilly or mountainous regions, leading to further destruction and blocking transportation routes.

2.1.9. *Preparedness and mitigation measures*

Mitigation measures for earthquakes hazard can be broadly divided into two categories: structural and nonstructural measures. Implementing a combination of these mitigation measures is crucial for reducing the impact of earthquakes on both buildings and communities.

2.1.9.1. *Structural mitigation measures for an earthquake*

1) *Building codes and standards*: the building codes and standards specified by BIS are enforced and regularly updated to ensure that structures are designed and constructed to withstand seismic forces in the future. Construction practices in seismic zones need to adhere to the seismic design codes. These codes provide guidelines for the construction of earthquake-resistant buildings and infrastructures, taking into account the seismicity of the region.

2) *Seismic retrofitting*: this includes retrofitting the already existing weak buildings and structures to improve their resistance to earthquakes, especially in regions with high earthquake risk. This includes activities like reinforcing foundations, adding shear walls and strengthening connections. In addition, following traditional earthquake-resistant structures can also help to build back in a better way.

3) *Infrastructure resilience*: this includes design buildings and construct critical infrastructure such as bridges, hospitals and power plants to be more resilient to seismic activity. Architectural and engineering inputs need to be put together like beams and columns to improve building design and construction practices. The structural design should be flexible and ductile, allowing them to bend and absorb seismic energy without collapsing during an earthquake.

4) *Base isolation*: this uses techniques to decouple buildings and infrastructure from the ground motion, reducing the transfer of seismic forces to the structure. Buildings should be made with a solid base and foundation.

2.1.9.2. *Nonstructural mitigation measures for earthquake*

1) *Land use planning*: this implements zoning regulations and land use planning which takes into account the seismic hazard to prevent or limit development in high-risk areas.

2) *Public awareness, education and preparedness*: this educates people through training programs and by conducting drills about the risks associated with earthquakes, and disseminating information related to preparedness and

safety measures are crucial components of reducing the risk and minimizing the impact of earthquakes in vulnerable regions.

3) *Risk assessment and monitoring*: these continuously assess earthquake risks and invest in monitoring and research to better understand seismic activity and improve prediction capabilities.

4) *Early warning systems*: these systems invest in early warning systems that can provide alerts and valuable seconds to take protective actions before the arrival of strong seismic waves.

5) *Emergency planning and preparedness*: this encourages people living in seismic hazard zones to adopt earthquake preparedness measures, including creating emergency plans, and being aware of evacuation routes. In addition, emergency plans that include procedures for evacuation, safe sheltering and post-earthquake response are developed and implemented.

6) *Critical facility protection*: this identifies and protect critical facilities, such as hospitals, emergency response centers and utilities, to ensure their continued operation during and after an earthquake.

2.2. Floods

2.2.1. *Meaning and concept*

Floods are natural disasters characterized by the overflowing of water onto land that is normally dry, resulting in widespread destruction and posing significant risks to human lives and infrastructure. They can occur from various factors including heavy rainfall, snowmelt, storm surges, ice jams, dam failure, human activities such as deforestation and urbanization, leading to different types of floods including flash floods, urban floods, coastal floods and riverine floods. The term ‘flood’ means inundation due to overflowing of a large volume of water, whatever its source. The impacts of floods are devastating, encompassing extensive infrastructure damage, loss of lives and livelihoods, health risks from contaminated water and economic hardship. Mitigation and preparedness measures such as floodplain zoning, early warning systems, infrastructure resilience, capacity building and community education are crucial for reducing the impact of floods and enhancing community resilience to tackle this natural hazard. There is no concept of flooding without large and unmanageable huge volume of water rising, swelling and overflowing onto the land. Flooding (in most cases) is a phenomenon concerning streams and rivers and their surrounding areas. Nevertheless, in the event of insufficient drainage, regions devoid of streams or rivers may still experience flooding. This condition is exacerbated by unplanned urban growth as well as obstructions in agricultural regions. The terms "floods" and "flooding" are frequently used interchangeably. Flooding is likely to occur in areas far from river or canal banks if there is an unexpected influx of water.

Furthermore, there are large bodies of water (besides streams and rivers) that could spill over to create flooding. Flooding or inundation is the overflowing by water of the normal confines of a river, stream, lake, sea or other body of water or accumulation of water due lack of drainage system over areas that are normally not submerged. Floods are the most frequent type of natural disasters characterized by the overflow of water from water bodies submerging the surrounding land areas and causing widespread damage to areas downstream that are typically dry. They can result from prolonged heavy rainfall, rapid melting of large amounts of snow, storm surges or dams or levees failure. Accumulation of large volumes of water poses significant risks to communities, including damage to property, and infrastructure, and loss of life. Low-lying areas, floodplains and regions with poor natural drainage are more susceptible to flooding, especially during heavy rainfall.

2.2.2. *Types of floods*

Floods can be classified depending on the source of water. Each type of flood presents unique challenges and requires strategies to minimize their impact on lives and infrastructure. Common types of floods are as follows:

1) *Riverine floods* or *river floods*: these floods result from prolonged and heavy rainfall or snow melt causing rivers and other water bodies to overflow their banks. They can lead to widespread damage to agricultural land, homes and infrastructure in the surrounding areas.

2) *Flash floods*: these occur suddenly and with little or no warning, often due to intense rainfall or cloud bursts over a short period or the rapid melting of snow or ice. These floods are known for their rapid onset and can be particularly dangerous due to the speed and force of the water.

3) *Coastal floods*: these floods are associated with cyclonic activities and due to high tides, storm surges along the coast, or tsunamis, leading to the

inundation and erosion of coastlines, damage to coastal infrastructure, and pose significant risks to human populations in low-lying coastal regions.

4) *Urban floods*: these types of floods are often associated with rapid urbanization, inadequate drainage systems and the covering of natural surfaces with impermeable materials and may lead to waterlogging, damage to urban infrastructure and disruptions to daily life in urban areas.

5) *Catastrophic floods*: these types of floods are triggered by extreme weather events such as hurricanes, typhoons or heavy monsoon rains and have the potential to cause widespread devastation, leading to loss of life and extensive damage to property and communities.

Figure 2.3. *Flooded Streets; Kashmir flood 2014. For a color version of this figure, see www.iste.co.uk/akbar/disastermanagement.zip*

2.2.3. *Causes of floods*

Floods can be caused by various factors, and they can result from both natural and human-induced processes. Here are some of the common causes of floods:

1) *Heavy rainfall*: intense and prolonged periods of rain can lead to flooding, especially in areas with poor drainage or in regions prone to monsoons or tropical storms. When rivers exceed their capacity due to heavy rainfall, snowmelt or a sudden influx of water from upstream, they can overflow their banks and cause flooding in adjacent areas.

2) *Coastal storm surges*: tropical storms, hurricanes or tsunamis can cause coastal flooding through storm surges, which are massive walls of water pushed ashore by strong winds.

3) *Climate change and snowmelt*: rapid melting of snow due to sudden temperature changes can overwhelm drainage systems and cause flooding, particularly in areas with poor water management. Rising global temperatures can lead to more intense and frequent extreme weather events, such as heavy rainfall and storms, increasing the likelihood of flooding.

4) *Urbanization and human activities*: as urban areas expand, natural land cover is replaced by impermeable surfaces like concrete and asphalt, leading to increased surface runoff and higher flood risk. Structural failures of dams or levees can release large volumes of water, resulting in catastrophic flooding downstream.

5) *Deforestation*: removing trees and vegetation can disrupt natural drainage patterns, leading to increased surface runoff and soil erosion, which can contribute to flooding.

2.2.4. Flood-prone regions of India

India has several flood-prone regions due to a combination of factors such as topography, climate and human activities. The geographical location of India renders it to be one of the most disaster-prone countries in the world. The most common flood-prone regions include Andhra Pradesh, Arunachal Pradesh, Assam, Bihar, Chhattisgarh, Delhi, Gujarat, Haryana, Himachal Pradesh, J&K, M.P., Maharashtra, Manipur, Meghalaya, Mizoram, Nagaland, Orissa, Pondicherry, Punjab, Rajasthan, Tamil Nadu, Tripura, Uttaranchal, U.P. and West Bengal. The most affected states are Assam, Bihar, Orissa, U.P. and West Bengal. Although the heavy rain resulting from the monsoons and the storm surge associated with cyclones also causes floods, the occurrence of floods in India is mainly associated with the river systems of the country. Floods are an annual feature in the Brahmaputra basin and cause heavy damage by inundation and erosion. The North-Western river basin, covering the states of Jammu and Kashmir, parts of Punjab, Himachal Pradesh and Rajasthan, has the river system of Jhelum, Sutlej, Beas, Ravi and Chenab. In Kashmir valley, the Jhelum is unable to carry the river discharge.

1) *Gangetic Plains*: the states of Bihar and Uttar Pradesh, which make up the Gangetic Plains, are particularly susceptible to flooding. Flooding frequently occurs in the areas around the Ganga and its tributaries, the Yamuna and Gandak, as a result of the Himalayan snowmelt and severe monsoon rains.

2) *Brahmaputra Valley*: the Brahmaputra River, flowing through Assam and parts of Arunachal Pradesh, is prone to frequent and severe flooding during the monsoon season. Heavy rainfall in the catchment areas, combined with glacial melt and snowmelt from the Himalayas, contributes to the floods in this region.

3) *Eastern Coastal Plains*: coastal areas in states like Odisha and West Bengal are susceptible to cyclone-induced floods and storm surges. Cyclones in the Bay of Bengal can bring heavy rainfall and cause riverine and coastal flooding.

4) *Western Ghats*: the Western Ghats, especially the states of Maharashtra, Karnataka and Kerala, experience flooding due to heavy rainfall during the monsoon season. The rivers originating in the Western Ghats, such as the Krishna and Godavari, can swell and cause floods.

5) *North-eastern states*: apart from the Brahmaputra Valley, other north-eastern states such as Meghalaya and Mizoram are prone to flooding, primarily due to their hilly terrain and heavy rainfall during the monsoons.

2.2.5. *Worst flood disasters in India*

Some worst-case studies of flood disasters in India include:

1) *Bihar flood, 2004*: the 2004 Bihar flood was one of the worst floods in Bihar, India in a decade. It affected 20 districts of Bihar and resulted in the loss of 885 human lives, with nearly 21.299 million people affected. The flood was caused by an alarming rise in water level due to heavy rainfall, which inundated fresh areas in several districts. The rivers Baghmati, Budhi Gandak, Kamla Balan, Adhwara, Kosi and Mahananda were flowing above the red mark at various places, with the Ganges crossing the danger mark for the first time at Farakka Barrage. This flood disaster had a huge impact on the region, leading to substantial loss of life and property (source: Relief Web – India: Situation report – Bihar floods, August 11, 2004).

2) *Mumbai flood, July 2005*: the Mumbai flood of 2005, also known as the July 26 or the Maharashtra flood, was caused by heavy rain. The city experienced approximately 944 mm of rain within 24 h, which paralyzed the entire city. It caused loss of lives and infrastructure. Over 1,000 people lost their lives and approximately 14,000 homes were destroyed. The floods caused significant economic disruption and infrastructural damage, with businesses and transportation severely affected.

3) *Assam flood, July 2012*: the Assam flood was caused by heavy rainfall and it caused multiple breaches in the Brahmaputra banks and its tributaries. Its impact was devastating and over 124 people, including 70 children, died, about 2.2 million people were affected and over 560 animals died, including 14 rhinos.

4) *Uttarakhand flood, June 2013*: the main cause of this year's flood was a cloudburst that brought plenty of rainfall, which in turn caused landslides and abrupt flash floods in the state's northern region. An estimated 4,094 people were killed and over 5,700 were missing (eventually presumed dead), making it one of the worst floods in India's recent history.

5) *Kashmir flood, September 2014*: the UT of J&K was hit by a devastating flood on September 7, 2015. As per studies, the main cause for this flood was heavy rainfall leading to massive floods that lasted for 8 days, and widespread damage to farms, roads, communication infrastructure, homes and hospitals. It was estimated that the damage had an economic value of 5,000–6,000 crores rupees.

6) *Chennai flood (2015)*: Chennai, Tamil Nadu was hit by a worse flood in 2015. The major cause for this flood was intensified low-pressure areas leading to huge amounts of rain in the coastal regions of Tamil Nadu and Andhra Pradesh. Massive flooding, with over 500 people presumed dead, more than 50,000 crore rupees in losses, and 60% of the city's power supply disconnected, leading to nonfunctional hospitals and submerged districts.

7) *Kerala flood (2018)*: the major cause for the Kerala flood was unprecedented rainfall during the monsoon season, combined with alleged delays in opening dam floodgates and man-made environmental factors. More than 400 reported deaths, lakhs displaced and estimated economic losses of 30,000 crore rupees, making it the worst flooding episode in Kerala in a century and causing an incredible amount of damage and destruction.

Figure 2.4. *Kashmir flood 2014. For a color version of this figure, see www.iste.co.uk/akbar/disastermanagement.zip*

2.2.6. *Impact of floods*

Floods are very dangerous and can have devastating effects on both the environment and human settlements. The impact of floods can be far-reaching and may include:

– *Environmental damage*: floodwaters can cause significant erosion of soil and riverbanks, leading to the loss of fertile land and disruption of ecosystems. Floods can also destroy the habitat of wildlife, leading to displacement and loss of biodiversity, impacting the overall ecological balance of affected areas. In addition, flood waters may carry pollutants and contaminants, leading to the contamination of drinking water and other water sources posing risks to public health and ecosystems.

– *Socioeconomic effects:* floods can lead to loss of life and cause injuries, particularly in areas where communities are unprepared or unable to evacuate in time. Floods can also cause damage to roads, bridges, buildings and other infrastructure, leading to disruptions in transportation and communication networks. The cost of repairing infrastructure, compensating for property damage and addressing the long-term impacts of floods can result in substantial economic losses for communities and governments.

– *Damage to agriculture*: agricultural lands can be inundated, leading to crop loss, soil erosion and damage to farm infrastructure. Floods can contaminate soil, and disrupt agricultural activities, leading to food shortages and economic losses.

– *Impact on plant life*: floodwaters can drown or uproot vegetation, disrupting plant communities and affecting the overall ecosystem balance.

– *Impact on human health*: floods can degrade water quality by carrying pollutants, debris and contaminants into water bodies. In addition, floods can create favorable conditions for the spread of waterborne diseases, posing significant health risks to affected populations.

2.2.7. *Mitigation measures for floods*

Mitigation measures include a combination of both structural and nonstructural measures that can significantly reduce the impact of floods on communities and ecosystems. By integrating these measures, it is possible to enhance flood resilience and minimize the adverse effects of flood disasters.

2.2.7.1. *Structural measures*

– *Physical infrastructure*: this involves constructing physical infrastructure to manage and mitigate the risk of floods, for example, building flood barriers, levees, dams and reservoirs to control and redirect floodwaters, and improving drainage systems to reduce the impact of floods, especially in urban areas. In addition, buildings and structures should be designed in such a way so that they can withstand and adapt to the impacts of climate change. Furthermore, it incorporates climate-resilient features in building designs, such as elevated foundations and flood-resistant materials are helpful structural measures to reduce the harmful effects of floods.

– *Flood barriers and embankments*: flood barriers and embankments are the physical structures that are built along riverbanks and coastlines to prevent floodwaters from inundating adjacent areas. Similar to barriers, flood walls can also protect rising water levels, particularly in urban areas.

– *Reservoirs and flood storage areas*: these structures are designed to retain excess water during heavy rainfall periods, reducing downstream flood risks.

– *Channelization*: this involves modifying rivers and streams to improve their flow capacity and reduce the risk of overflow.

– *Storm surge barriers*: coastal defenses are designed to protect against tidal surges and coastal flooding, particularly in low-lying areas and coastal regions.

– *Flood diversion channels*: these include construction of diversion channels to redirect excess water away from populated areas to less vulnerable regions.

– *Stormwater management infrastructure*: this includes developing stormwater management systems, including retention ponds, drainage channels and permeable surfaces, to control urban flooding caused by heavy rainfall especially in urban areas.

2.2.7.2. *Nonstructural measures*

These measures are crucial for reducing the harmful impacts of foods and may include:

– *Flood forecasting early warning systems*: this is the most important mitigation measure and involves utilizing the technology and communication networks to provide advance notice to the people about impending flood disaster, allowing them for timely evacuation and preparation.

– *Ecosystem-based approaches*: this involves the utilization of natural ecosystems to reduce the impact of floods and enhance resilience of people. This includes restoring wetlands, maintaining natural drainage systems and preserving floodplains to absorb and retain floodwaters.

– *Floodplain management and zoning*: this involves regulation of land use in flood-prone areas to minimize exposure associated with flood risk.

– *Flood insurance and risk transfer mechanisms*: flood insurance and transfer of risk involves provision of financial protection to individuals, property and businesses against flood-related losses.

– *Land use planning*: land use planning involves implementation of regulations and guidelines for constructing new buildings and infrastructure that are more flood resistant.

– *Public awareness and education*: education and awareness are important for informing communities about flood risks, preparedness and response strategies. Moreover, people should be encouraged to construct engineered structures to enhance resilience.

– *International cooperation*: facilitating collaboration and information exchange between countries to address transboundary flood risks, including sharing data and best practices, and developing joint early warning systems for river basins that traverse multiple countries.

So, we can say, effective flood management involves a combination of both structural and nonstructural measures, as each type of measure has its own strengths and limitations. An integrated approach that combines physical infrastructure with policy, planning and community engagement is essential for comprehensive flood risk reduction and management.

2.3. Landslides

2.3.1. *Meaning and concept*

These are defined as the mass movement of rock debris or mud down a slope and have come to include a broad range of motions, whereby falling, sliding and flowing under the influence of gravity. A serious risk to the environment, infrastructure and human settlements is posed by landslides, often known as "landslips". They also occur as secondary effects of heavy rainstorms, cloudbursts, flash floods and earthquakes. The word "slide" refers to the downslope movement of geologic materials that primarily occurs on thin zones of extreme shear strain or surfaces of rupture. They usually occur as a result of changes, either sudden or gradual, in the structure and composition of rocks or as a result of slope failures. In most cases, abrupt or gradual changes in the composition and structure of rocks or vegetation on a slope result in landslides. They are also triggered by natural events such as intense rain, earthquakes, volcanic eruptions, human activities, poor ground conditions, geomorphic phenomena or undercutting of the base of slopes by rivers. Heavy rainfall spells combined with inadequate drainage systems frequently result in the movement of rocks, soil, earth and debris down the slopes or hills. It is the most often used and generally recognized collective word for the majority of massive slope movements. These incidents carry a risk to human safety and can result in serious damage to infrastructure, property and the environment. Among all of the natural calamities that affect mountainous areas, landslides are one of the most concerning. They are one of the biggest and most common natural hazards in India, which cause significant fatalities as well as harm to human habitations, agricultural areas, forests and communication routes. The Himalayan subcontinent, which includes the Northeastern mountain ranges, is most vulnerable to landslide hazards due to its complex physiographic, seismo-tectonic and climatological characteristics. The Vindhyas, Nilgiris and a portion of the Western Ghats are also at risk and this usually occurs due to the torrential monsoon. Landsliding is also caused by the toe erosion and removal of vegetation. The communities located at the base of steep slopes and at the mouth of streams that emerge from mountain valleys, communication lines, underground utilities and roads are more vulnerable to landslides. Slope alteration and human intervention have increased this effect.

Landslides occur in almost all of the hill ranges of India that encompass a broad range of land formations and processes involving the movement of soil and rock downslope under the impact of gravity. Since the Himalayas are a relatively young mountain range with fragile and weak rock structure, it has been noted that they are the locations of the most common landslide events. Landslides cause significant damage to farms, plantations, homes, roads and other structures. The debris originating from landslides results in the siltation of streams, lakes and rivers, inducing further problems of erosion and floods. Damage to structures, loss of lives and properties occur extensively every year as a result of landslides.

Figure 2.5. *Landslides. For a color version of this figure, see www.iste.co.uk/akbar/disastermanagement.zip*

2.3.2. *Types of landslides*

Over the years, different systems of landslide classification have evolved based on the mode and the rate of the movements, type and nature of the material involved, deposits' age as well as additional elements like the prevailing weather. The events of landslides occur in a variety of ways, each with unique characteristics and reasons. Comprehending the particular kind of landslide is essential for evaluating possible hazards and putting in place efficient preventative measures. Every kind has a unique set of traits and behaviors that affect how it affects the situation and what needs to be done to manage the risks involved. Here are some of the most common types of landslides:

1) *Rock falls*: rock falls occur when rocks or boulders detach from a steep slope or when the rapid free-fall of rock fragments from a steep slope. These types of landslides are often triggered by many factors such as **weathering** and erosion of the rock face, human activities, as well as by seismic activity.

2) *Debris flows*: debris flows, sometimes referred to as mudflows, happen when water, rock and soil rush down a slope quickly. These kinds of landslides are frequently brought on by sudden or intense snowmelt or rain, which can saturate the ground and cause the slope to become unstable. These are fast-moving, liquefied flows of saturated debris which usually contain a high percentage of silt and clay. These are fast-moving, liquefied landslides consisting of a mixture of water, rock, soil and other debris. They usually appear in regions with steep slopes where heavy precipitation and quick snowmelt take place.

3) *Debris slide* or *debris avalanche*: this is a type of fast-moving landslide that involves the downward movement of a mixture of materials like rock, soil, vegetation and other debris due to heavy rainfall, earthquakes, volcanic eruptions or human activities that destabilize the slope of a landmass. It is rapid movement and can travel long distances at high speed and cause huge damage and destruction.

4) *Landslides*: these occur when a mass of soil or rock slides down a slope due to gravity. These types of landslides can be triggered by a variety of factors, including heavy rainfall, seismic activity and human activities such as construction or mining activities.

5) *Creep*: this is a slow, continuous movement of soil or rock down a slope. This kind of landslide is frequently caused by temperature and moisture variations in the soil, as well as long-term weathering and erosion of the slope. It is typically not as sudden as other types of landslides and is often driven by the force of gravity.

6) *Slumps*: these occur when a block of soil or rock rotates and slides down a curved slope. These types of landslides are often triggered by a combination of geological factors, such as the type of soil or rock and the slope angle. Slumps are rotational landslides involving the slipping of several rock blocks with a backward movement w.r.t slope where the descending mass moves along a surface, resulting in a characteristic spoon-shaped landform. They are commonly triggered by saturation of the ground from prolonged rainfall.

2.3.3. *Causes of landslides*

Landslides occur when the stability of a slope is compromised, leading to the downward movement of material. Some of the common causes of landslide hazards are as follows:

1) *Geological factors*: various types of rocks, such as shale, are more prone to weathering and erosion resulting in the undercutting of slopes by running water increases the likelihood of landslides. Steep slopes are more vulnerable to landslides as gravitational forces can easily overcome the strength of the materials. The presence of weak layers, such as clay or silt, within the slope can increase the risk of landslides.

2) *Weather and climate*: sudden or prolonged periods of intense rainfall can saturate the soil and reduce its stability, leading to landslides. An increase in temperature and rapid melting of snow in summer or monsoon season can saturate the ground and trigger landslides. Additionally, alternating freezing and thawing processes of water within the soil or rock can weaken the materials, which in turn lead to landslides.

3) *Human activities*: anthropogenic activities or human activities like deforestation, mining activities and construction works alter the natural stability of slopes. Removal of vegetation and overgrazing make fragile slopes unstable and reduce soil cohesion, making them more susceptible to

landslides. Also, due to developmental activities such as slope cutting for the construction of roads, widening of highways, industries, excavation, urbanization, blasting, and use of heavy machinery changes in land use and unsustainable agricultural practices can change the natural stability of slopes, increasing the risk of landslides. In addition, poorly managed irrigation and leaking pipes can lead to increased water infiltration, which can destabilize the slopes.

4) *Earthquakes and volcanic activity*: seismic events like earthquakes can contribute to landslides by pushing soil materials down particularly on steep slopes, causing unstable slopes and triggering landslides. In addition, volcanic activities have potential to cause landslides and can lead to the rapid movement of debris due to volcanic explosions.

Ground causes:
1) Weak, sensitive or weathered materials 2) Adverse ground structure (joints and fissures) 3) Physical property variation (permeability and plasticity)
Morphological causes: 1) Ground uplift (volcanic, tectonic) 2) Erosion (wind, water) 3) Deposition loading in the slope crest 4) Vegetation removal (by forest fires, droughts)
Physical causes: 1) Intense or prolonged precipitation 2) Rapid drawdown 3) Earthquake 4) Volcanic eruption 5) Thawing 6) Shrink and swell 7) Artesian pressures
Man-made causes: 1) Excavation (particularly at the toe of slope) 2) Construction and urbanization 3) Uncontrolled grazing 4) Deforestation 5) Irrigation and water leakage 6) Mining and quarrying 7) Blasting activities and artificial vibration

Table 2.2. *Causes of landslides*

2.3.4. *Landslide-prone regions of India*

India is prone to landslides due to its diverse topography, geological formations and climatic conditions. Some of the landslide-prone regions in India include:

– *The Himalayan region*: the Himalayas, especially in states like Uttarakhand, Himachal Pradesh and Jammu and Kashmir, are highly susceptible to landslides due to steep slopes and fragile geological formations, particularly during the monsoon season and after seismic events.

– *Eastern and Western Ghats*: certain areas in the Eastern Ghats, including parts of Odisha, Andhra Pradesh and Tamil Nadu, are susceptible to landslides, particularly in hilly and forested regions. The Western Ghats, particularly in states like Kerala, Karnataka and Maharashtra, are also prone to landslides, especially during the monsoon season, due to heavy rainfall and deforestation.

– *North-eastern states*: states in the north-eastern region of India, such as Meghalaya, Assam, Arunachal Pradesh and Nagaland, are prone to landslides due to their hilly terrain, intense monsoons and seismic activity.

– *Central India*: certain hilly areas in central India, including parts of Madhya Pradesh, Chhattisgarh and Maharashtra, are also susceptible to landslides, especially in areas with deforestation and improper land use.

2.3.5. *Effects of landslides*

Landslides can have a wide range of significant impacts on the environment, infrastructure and communities. Some of the major impacts of landslides are as follows:

1) *Loss of life and causalities*: those communities who are living at the foothills and mountainous regions with inadequate infrastructure are at greater risk of landslides leading to loss of life and injuries particularly in densely populated areas.

2) *Disruption of infrastructure and transportation*: landslides can cause extensive damage to roads, bridges, buildings and utility lines, disrupting transportation, communication and essential services.

3) *Environmental degradation*: landslides can result in significant environmental degradation, including the destruction of habitat and ecosystems, loss of biodiversity and increased sedimentation in water bodies, which can have long-term consequences for the environment.

4) *Economic impact and loss of livelihood*: the economic costs of landslides can be substantial, including expenses related to rescue and relief operations, rebuilding infrastructure and addressing long-term social and economic impacts. In addition, agricultural land, cash crops, forests and other natural resources can be destroyed or rendered unusable by landslides, leading to the loss of livelihoods for communities that depend on these resources.

5) *Psychosocial impact*: the trauma and stress resulting from the loss of life, property and livelihoods can have long-lasting psychosocial impacts on individuals and communities affected by the landslides.

2.3.6. *Mitigation measures of landslides*

Natural hazards especially landslides can cause significant damage and destruction to human life, infrastructure and the environment. Implementation of effective mitigation measures is very important for reducing the potential risks associated with landslides. Here are some important mitigation measures that can be taken to reduce the harmful impact of landslides.

2.3.6.1. *Structural measures*

Some of the preventive or mitigation measures to be taken to reduce the risk of landslides are as follows:

1) *Rock fall barriers and slope stabilization*: this includes installing rock fall barriers made up of steel cables, wire mesh, rock bolts or other materials to intercept and prevent falling rocks and debris, preventing them from causing landslides downslope.

2) *Engineering solutions*: engineering solutions such as retaining walls, slope stabilization, rock bolts and drainage systems to reinforce unstable slopes can help stabilize the soil to prevent downslope movement and

reduces the risk of landslides. The use of engineering techniques drains water away from slopes to reduce saturation and erosion.

3) *Infrastructure design and retrofitting*: this incorporates landslide-resistant designs and materials into infrastructure projects, such as bridges, roads and buildings, especially in landslide-prone regions. In addition, it retrofits existing infrastructure to improve resilience to landslide impacts.

4) *Terracing and diversion structures*: this includes creating terraces or step-like features on steep slopes and will reduce the force and overall angle of the slope and minimize the risk of mass movement. In addition, diversion structures will also help in diverting the movement of mass during landslides from slopes and reducing the risk of landslide initiation.

5) *Bioengineering techniques*: these include implementing vegetation-based measures, such as planting trees, shrubs and grasses, to stabilize slopes. These will act as a natural barrier to reduce erosion.

2.3.6.2. *Nonstructural mitigation measures*

1) *Land-use planning and zoning*: this includes identifying vulnerable regions and avoiding high-risk areas for development and regulates construction in landslide-prone areas by following zoning laws and building codes and promoting safer construction practices for sustainable development.

2) *Risk assessment and mapping*: this conducts detailed risk assessments and mapping of landslide-prone areas to identify high-risk zones and inform land-use decisions and infrastructure development. Creating hazard maps help in the identification of hazard zones and vulnerable areas of landslides. So, by having such maps in hand will help people to restrict construction and developmental activities in these areas. This will also help in land use planning.

3) *Public awareness and education*: this includes conducting public awareness campaigns and educational programs to inform communities about the risks of landslides and emergency procedures and provide training to local authorities, emergency responders and communities on preparedness and response measures for landslides.

4) *Early warning systems and monitoring*: this includes establishing early warning systems that utilize technology such as monitoring sensors and weather forecasts to alert people and communities to provide advance notice of potential landslide hazards for timely evacuation and disaster risk reduction. In addition, it installs and maintains landslide monitoring systems, including inclinometers, ground-penetrating radar and remote sensing and geo-informatics technologies, to detect slope movements.

5) *Afforestation and erosion control*: this includes implementing measures to preserve and restore natural vegetation, and managing drainage system by controlling erosion to stabilize slopes reduces landslide risk. In addition, it also includes conducting reforestation and afforestation programs in landslide-prone areas at different levels to enhance slope stability.

6) *Regulation and policy*: this includes developing and enforcing regulations and policies to ensure responsible land use and development practices in landslide-prone areas. It integrates landslide risk assessment and environmental impact assessments into land-use planning.

2.3.7. *Safety tips*

Landslides can usually pose significant risks and threat to people and their property. Understanding the dos and don'ts, as well as adhering to safety tips, can help individuals and communities mitigate the impact of landslides and ensure their safety.

2.3.7.1. *Dos*

1) Always look out for features indicating deterioration or movements on slopes like tension cracks, erosion gullies on slopes or toe cutting at the slopes.

2) Stay updated and informed on weather forecasts and landslide warnings issued by local authorities.

3) If authorities issue evacuation orders or advisories, follow their instructions promptly and evacuate to safer areas.

4) Keep important and valuable documents, emergency supplies, in an easily accessible location to take with you if evacuation becomes necessary.

5) Look out for any stress in adjoining areas.

6) Study the drainage pattern particularly blockings and changes in drainage channels.

7) Try to collect as much information as possible on failure plane/features showing distress.

8) Try to correlate the features with ambient field conditions.

2.3.7.2. *Don'ts*

1) Avoid cutting at the toe portion due to excavations/streams.

2) Avoid concentrated runoff through the body of slope.

3) Avoid percolation of water through cracks and broken drains.

4) Avoid cutting/grazing of plants and bushes.

5) Do not ignore warnings: take landslide warnings seriously and avoid underestimating the potential risks associated with unstable slopes and terrain.

6) Avoid waiting until the last minute to evacuate if landslide risks are imminent. Proactive action is crucial for personal safety.

7) Refrain from entering areas identified as at high risk for landslides, especially during or after heavy rainfall or seismic activity.

8) Avoid constructing buildings or homes in areas prone to landslides, and if possible, relocate to safer locations.

9) Do not assume safety after a landslide: exercise caution even after a landslide has occurred, as subsequent landslides can still pose risks due to destabilized terrain.

2.4. Cyclones

2.4.1. *Introduction to cyclones*

Among the extreme weather events, cyclones are by far the most devastating, both by causing loss of human lives as well as economic losses. These are also called tropical cyclones because they form over the warm waters of tropical oceans. Intense low-pressure areas forming over higher latitudes either over the land or over the ocean are known as extra tropical cyclone. Hazards associated with tropical cyclone result from very strong winds causing damages of various types: very heavy rains cause widespread floods and storm surges. Storm surge is the sudden rise of sea level caused by cyclone. Cyclones are meteorological hazards characterized by low-pressure centers, strong winds and thunderstorms. Wind speeds can vary within the storm, with the highest speeds occurring in the eyewall. These are some of the biggest threats to life and property. The Greek word "cyclos", which means snake coils, is from where the term "cyclone" originates (Henry Paddington). The entire lifecycle of a cyclone may last from a few days to a few weeks with a diameter ranging from a few tens of kilometers to several hundred kilometers, at altitudes of 10–15 km (6–9 miles). They occur mainly in low-pressure areas. There is hardly any year when India is not visited by a severe cyclone (also called a cyclonic storm or tropical cyclone). The Bay of Bengal and the Arabian Sea are two regions known for being prone to cyclones and more cyclones occur in the Bay of Bengal than in the Arabian Sea, with the ratio being 4:1 approximately. The cyclones are most common in pre-monsoon (April–May) and post monsoon (October–December) seasons. Cyclones, also known as hurricanes or typhoons in different regions of the world, are powerful storm systems characterized by strong winds, heavy rainfall and storm surges. They occur clockwise in the southern hemisphere and anticlockwise in the northern hemisphere. These intense tropical storms can cause widespread destruction and pose significant risks to human life and infrastructure. At the center of the storm, there is an eye of cyclone which is relatively calm area with light winds and clear skies and is characterized by sinking air. The area surrounding the eye, where the most intense winds and heavy rainfall occur during a cyclone is known as eye wall. Cyclones are characterized by a low-pressure center at the surface. This low pressure is due to the warm air at the surface rising and creating a vertically stacked system. The size of a cyclone contributes to its ability to influence weather patterns over extensive regions. They are vertically stacked structure with strong convection extending from the surface to the upper troposphere. The vertical structure allows for the efficient transfer of heat and moisture, contributing to the storm's development and maintenance. Cyclones typically have seasonal patterns and occur during particular times of the year.

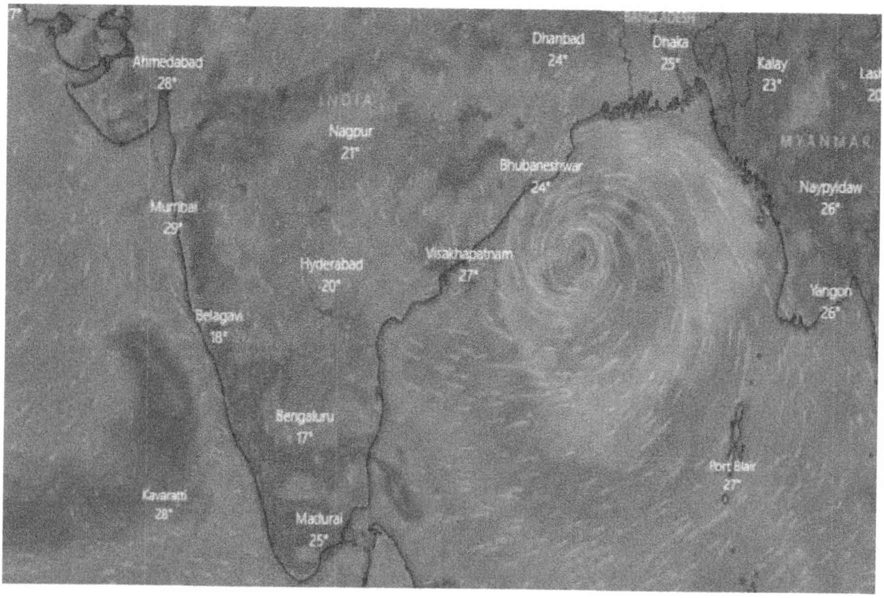

Figure 2.6. *Cyclones. For a color version of this figure, see www.iste.co.uk/akbar/disastermanagement.zip*

2.4.2. *Types of cyclones*

1) *Extratropical cyclones* or *temperate cyclones*: these are cyclones forming outside the tropics and are often associated with mid-latitude weather systems. They can bring a variety of weather conditions, including rain, snow and strong winds.

2) *Polar cyclones*: cyclones that occur near the polar regions, such as the Arctic or Antarctic. These systems are usually smaller in scale compared to tropical cyclones.

3) *Subtropical cyclones*: cyclones that have characteristics of both tropical and extratropical cyclones. They can form over subtropical waters and have a less symmetric wind field compared to tropical cyclones.

2.4.3. *Conditions for formation of cyclones*

Cyclones typically form over warm ocean waters near the equator. The basic ingredients for their formation include the following:

– *Warm ocean water*: cyclones form over warm ocean waters, typically with sea surface temperatures of at least 26.5°C (about 80° F) or warmer. The warm water provides the necessary heat and moisture to fuel the storm's development.

– *Moisture*: high levels of humidity in the lower to mid-levels of the troposphere provide the necessary moisture for cyclone formation. Adequate moisture in the atmosphere is necessary for the condensation of water vapour, releasing latent heat that further warms the surrounding air and promotes the storm's growth.

– *Coriolis effect*: the rotation of the Earth causes the developing storm system to spin, leading to the characteristic cyclonic circulation. The formation of cyclones requires specific conditions, and these systems typically develop over warm ocean waters. The Coriolis effect, caused by the rotation of the Earth, is important for the formation of cyclone. Cyclones cannot form within about 5° of the equator, where the Coriolis effect is too weak. As air converges toward the low-pressure center, the Coriolis effect deflects it, leading to the development of a rotating system.

– *Depth of warm water*: the warm water layer must extend to a sufficient depth (usually around 50 m or 164 feet) to provide a continuous supply of heat to the developing cyclone. This depth ensures sustained energy transfer to the atmosphere.

– *Preexisting disturbance*: cyclones often form from preexisting weather disturbances, such as tropical waves or low-pressure systems. These disturbances provide the initial conditions for cyclone development.

2.4.4. *Cyclone-prone regions*

India has a long coastline of 7,516 km, a major portion of which is exposed to tropical cyclones. Certain coastal and low-lying areas are particularly vulnerable to the impacts of cyclones due to their geographical location and environmental conditions. These hazards are more likely to occur in specific regions of the world where the necessary conditions for their formation are met. The primary cyclone-prone regions are as follows:

– *North Atlantic Ocean*: the North Atlantic Basin, particularly the tropical North Atlantic, is a well-known breeding ground for hurricanes. The season for Atlantic hurricanes begins on June 1 and ends on November 30.

– *North Pacific Ocean*: the Northwest Pacific Basin is the most active tropical cyclone basin on Earth. Typhoons (the term for cyclones in the Northwest Pacific) can form throughout the year, but the peak season is from May to October.

– *Southwest Pacific Ocean*: the Southwest Pacific Basin, including the South Pacific and Australian regions, experiences tropical cyclones. The cyclone season in the South Pacific typically runs from November to April.

– *Indian Ocean*: the Indian Ocean, particularly the northern part, is prone to tropical cyclones. The Bay of Bengal and the Arabian Sea are known cyclone breeding grounds. The cyclone season in the northern Indian Ocean usually spans from April to December. The region around Madagascar and the southwestern Indian Ocean is also susceptible to tropical cyclones. The cyclone season in the Southwest Indian Ocean typically occurs from November to April.

2.4.5. *Recent case studies of cyclones*

Several major cyclones have left a significant impact on various regions around the world. Here are a few case studies of noteworthy cyclones:

– *Orissa Super Cyclone (1999)*: October 1999 was a particularly unfortunate month for the cyclone-prone state of Orrisa. The severely damaged state had not yet finished assessing the harm caused by the extremely powerful hurricane that struck Ganjam region on October 17, 1999, with gusts of up to 180 kmph, when another cyclonic storm hit a wide area of the Orissa coast on October 29, 1999 with super cyclonic intensity seriously affecting 15 prosperous districts of the state. Orissa super cyclone of 1999 was the strongest cyclonic storm to strike India in the 20th century.

– *Hurricane Katrina (2005)*: one of the most devastating hurricanes in U.S. history, Katrina made landfall on the Gulf Coast, particularly affecting New Orleans. The storm surge breached levees, leading to widespread flooding. The response and recovery efforts faced criticism for their shortcomings, and the hurricane highlighted the vulnerability of coastal cities to extreme weather events.

– *Cyclone Nargis (2008)*: Nargis struck Myanmar (Burma) in May 2008, causing catastrophic damage. The storm's powerful winds and storm surge led to extensive flooding, particularly in the Irrawaddy Delta. Nargis resulted in a significant loss of life and highlighted the challenges of disaster response in politically isolated regions.

– *Super Typhoon Haiyan (Yolanda) (2013)*: Haiyan was one of the strongest tropical cyclones ever recorded. It struck the Philippines with powerful winds and a massive storm surge, causing widespread destruction and a significant loss of life. The typhoon underscored the vulnerability of coastal communities to extreme weather events and prompted discussions on climate change and adaptation.

– *Cyclone Hudhud (2014)*: Cyclone Hudhud was a powerful tropical cyclone that struck the eastern coast of India in October 2014. It was the strongest cyclone of the North Indian Ocean in terms of wind speed. The cyclone rapidly intensified over the warm waters of the Bay of Bengal, reaching its peak intensity on October 10, 2014, and made landfall near the city of Visakhapatnam in the Indian state of Andhra Pradesh on October 12, 2014. The strong winds, with speeds reaching up to 175 km/h (109 mph), uprooted trees, damaged buildings and disrupted power and communication lines. The total economic losses were estimated to be in the billions of dollars.

– *Cyclone Fani (2019)*: Cyclone Fani made landfall in India on May 3, 2019. It was an extremely severe cyclonic storm that affected the eastern coast of India, particularly the states of Odisha, West Bengal and parts of Bangladesh. Fani was one of the strongest cyclones to hit the Indian subcontinent in recent years, causing significant damage to infrastructure, displacing people and leading to loss of life. The storm originated from a low-pressure area in the Bay of Bengal and intensified into a powerful cyclone before making landfall near Puri in Odisha.

– *Cyclone Amphan (2020)*: Amphan became one of the strongest cyclones recorded in the Bay of Bengal. It struck the India-Bangladesh border, causing storm surges, heavy rainfall and wind damage. Amphan highlighted the importance of early warning systems and preparedness measures in densely populated coastal areas.

– *Cyclone Tauktae (2021)*: Cyclone Tauktae was a powerful tropical cyclone that affected multiple countries in the Arabian Sea region in May 2021. The storm originated from a low-pressure area in the Arabian Sea and rapidly intensified into a very severe cyclonic storm. Cyclone Tauktae made landfall on the western coast of India in the state of Gujarat on May 17,

2021. Before reaching Gujarat, the cyclone had brushed past the western coast, affecting states like Kerala, Karnataka, Goa and Maharashtra.

2.4.6. *Effects of cyclones*

The effects of cyclones are devastation and can lead to loss of life, injury and trauma for individuals and communities. Due to high-speed winds, cyclones are known to cause extensive damage to buildings, trees and infrastructure. They eradicate the structures and the people caught in them. Torrential rains due to cyclone cause major damage and can lead to inland flooding, landslides, soil erosion and mudslides leading to serious problem to the people who become shelterless. The combination of high winds and low atmospheric pressure can cause an abnormal rise in the sea level, as a result sea water can cause coastal flooding and erosion. Cyclones can also disrupt marine ecosystems, causing damage to coral reefs and affecting fish populations. Tropical cyclones are associated with potential risks and are the biggest threats to humans and their property. They include other individual hazards associated with it such as tornadoes and the disruption of essential services like power, water and transportation. In most cases, they are followed by torrential downpours, flash floods in coastal areas, storm surges that travel great lengths inland, up to 15 km, on top of tidal waves causing damage or even demolishing structures and roofs. Water pollution by flooding can result in water-borne diseases like malaria, diarrhea and virus epidemics. Floodwaters can contaminate piped and ground water supplies. Strong winds and precipitation will destroy any crops and food supplies situated in places with low elevation. Salt extracted from salt from the sea water may be deposited on the agricultural land and increase the salinity. The loss of the crop may lead to acute food shortage that could result from the crop's loss. Strong winds damage telephone and electrical lines, satellite dishes, broadcasting services and communication towers. Road and rail transportation links may also be disrupted.

2.4.7. *Mitigation and preparedness*

For minimizing the effects of cyclones, some measures and policy matters cannot be initiated by the community itself but needs intervention of the government. Complete prevention of cyclones is not possible but definite structural and nonstructural measures of long-term nature can be undertaken to mitigate the suffering of cyclone-affected people. Efforts to mitigate the impact of cyclones and enhance preparedness include two important measures: structural and nonstructural measures.

2.4.7.1. *Structural measures*

Structural mitigation measures for cyclones involve designing and constructing buildings and infrastructure in a way that reduces vulnerability to the impacts of cyclonic winds, storm surges and heavy rainfall. These measures aim to enhance the resilience of structures and communities. They include implementing and enforcing construction standards to enhance the resilience of buildings and infrastructure to cyclone impacts. These codes should address factors such as wind loads, structural design and the use of durable materials. Cyclone-resistant structures, including homes, schools, and critical infrastructure, withstand the wind forces associated with cyclones. This may involve reinforced concrete construction, the use of storm-resistant roofing materials and proper anchoring of structures. For example, installing roofing materials made to resist strong winds, such as reinforced roofing systems, and securing roof structures to the building frame.

– *Elevated foundations*: these include elevating buildings and critical infrastructure above expected storm surge levels. This helps prevent flooding and reduces the risk of damage to structures and their contents.

– *Storm surge barriers*: these include implementing coastal defenses such as storm surge barriers, levees and seawalls to reduce the impact of storm surges on coastal communities. These structures can help prevent or minimize flooding.

– *Proper drainage systems*: these systems design and maintain effective drainage systems to prevent flooding caused by heavy rainfall. Adequate drainage helps reduce the risk of structural damage and minimizes the potential for water-related hazards.

– *Building retrofitting*: this retrofits existing structures to meet current cyclone-resistant standards. This may involve strengthening building components, such as roofs and walls, to improve overall resilience.

– *Community infrastructure resilience*: this includes to plan and construct critical community infrastructure, such as hospitals, schools and emergency shelters, with a focus on resilience to cyclone impacts. These facilities should be designed to remain functional during and after a cyclone.

2.4.7.2. *Nonstructural measures*

Nonstructural measures include the following:

– *Forecasting and early warning systems*: cyclone forecasts are provided by warning centers by utilizing meteorological and forecasting data to provide advance warnings of cyclone events. They include establishing and maintaining effective early warning systems to provide timely and accurate information about impending cyclones. This includes meteorological monitoring, communication systems and community alerts.

– *Preparedness, response and evacuation plans*: these include developing and regularly updating emergency preparedness and response plans at local, regional and national levels. These plans should outline the roles and responsibilities of different agencies and stakeholders during a cyclone event. They also include involvement of local communities in the development of cyclone preparedness and response plans. These include identifying safe evacuation routes, establishing emergency shelters and organizing community drills.

– *Community education and awareness*: creation of proper awareness about cyclone risks and preparedness measures within vulnerable communities and people should be encouraged to structural mitigation measures for cyclones that involve designing and constructing buildings and infrastructure in a way that reduces vulnerability to the impacts of cyclonic winds, storm surges and heavy rainfall. These measures aim to enhance the resilience of structures and communities.

– *Land use planning and policies*: this includes implementing land use planning regulations to control development in high-risk areas. These include zoning restrictions and building codes to ensure that new constructions are resilient to cyclones and located in safer zones. They also include enacting and enforcing policies and regulations that promote resilient construction practices, safe building codes and land use planning. This involves creating a legal framework that encourages cyclone-resistant infrastructure.

– *Insurance and risk transfer*: this includes encouraging the use of insurance mechanisms and risk transfer strategies to mitigate the financial impact of cyclone damage. This can include the development of insurance products tailored to the specific needs of cyclone-prone regions.

– *Capacity building*: this includes building the capacity of local authorities, community leaders and emergency response teams to effectively manage and respond to cyclone events. This includes training programs, workshops and the provision of necessary resources.

– *Hazard mapping*: A hazard map will show where the cyclone is most likely to occur. The map is prepared with data input of past climatological records, history of wind speed and frequency of flooding.

– *Environmental conservation*: these measures include preservation of mangroves and natural barriers, which can help reduce the impact of storm surges and provide a buffer against cyclone-related damage.

2.4.8. *Cyclone warning system*

A cyclone warning system is a critical component of disaster management that aims to provide timely and accurate information about approaching cyclones, enabling individuals, communities and authorities to take necessary precautions and actions to minimize the impact of the cyclone. The system typically involves the use of advanced meteorological technology, communication networks and public outreach efforts to disseminate warnings and advisories. In regions prone to cyclones, early warning systems play a pivotal role in mitigating the devastating impacts of these powerful storms. Cyclone warnings are issued by meteorological agencies based on advanced weather monitoring technologies and predictive models. These warnings are disseminated through various communication channels, including radio, television, social media and mobile phone alerts, to reach as many people as possible in the affected areas. The warnings typically include information about the cyclone's expected path, intensity and potential hazards such as strong winds, heavy rainfall, storm surges and recommendations for evacuation and safety measures. Timely and accurate cyclone warnings allow authorities to initiate evacuation procedures, mobilize resources for emergency response and enable communities to prepare by securing their homes and belongings. By providing early warnings, communities can better prepare and respond to cyclones, ultimately saving lives and minimizing the destruction caused by these natural disasters.

Cyclone warnings are disseminated by the following means: (i) telegrams with the highest priority; (ii) telecast through Doordarshan; (iii) broadcast through AIR; (iv) press bulletins; (v) broadcasts through P&T's coastal radio stations for ships in the high seas and coastal waters; (vi) through INSAT-based Cyclone Warning Dissemination Systems (CWDS); and (vii) through Global Maritime Distress Safety System (GMDSS) using the International Maritime Satellite (INMARSAT). The present cyclone surveillance system in the country is such that no cyclone in the region can escape detection at any time of its life period.

The important components of cyclone warnings are the forecast of the future path, intensity and the associated destructive weather such as strong winds, heavy rainfall and storm surge. For the forecast of path of the cyclone and storm surges, the modern methods, which utilize computers, are applied in addition to conventional methods. The forecasting techniques used in India are comparable to other advanced countries in the world. The primary importance of a cyclone warning system is to reduce the risk to human life and property by providing early and accurate warnings of impending cyclones, allowing for timely evacuation and preparation. It is of paramount importance in mitigating the impact of cyclones, protecting lives and livelihoods, and fostering community resilience in cyclone-prone regions.

Figure 2.7. *Safety tips for hurricanes*

2.5. Tsunamis

2.5.1. *Meaning and concept*

Figure 2.8. *A tsunami. For a color version of this figure, see www.iste.co.uk/akbar/disastermanagement.zip*

A tsunami is a sequence of extremely long ocean waves, usually extending hundreds of kilometers, that are brought on by significant ocean disturbances such underwater earthquakes, volcanic eruptions, landslides and other geological phenomena. Tsunami is a Japanese word meaning “Harbor wave”, represented by two letters “Tsu”, means harbor, and “nami”, means “wave”. Tsunamis are sometimes referred to as seismic sea waves due to their seismic origin. These disturbances displace a rapid or massive volume of water, resulting in the propagation of tsunami waves across the ocean posing significant hazards to coastal regions. Tsunamis have long wavelengths and can travel across the ocean at speeds exceeding 800 km/h, making them difficult to detect and allowing them to propagate over vast distances, while at sea tsunamis may not be noticeable due to their long periods (the time between successive waves), but as they approach shallower coastal waters, their energy becomes compressed, resulting in a significant increase in wave height. World Tsunami Awareness Day is observed every year on November 5 to increase public awareness of the risks posed by tsunamis on social media as well as through other platforms and to prepare communities for resilience as well as the importance of early warning systems to protect people and prevent disasters. This also serves as a day to remember past catastrophic tsunamis, such as the Indian Ocean tsunami of 2004, and draw lessons from them to better prepare for future calamities.

2.5.2. *Causes of tsunamis*

The main causes of tsunamis are as follows:

1) *Underwater earthquakes*: the most common cause of tsunamis is the vertical displacement of the seafloor during underwater earthquakes. When the seafloor shifts vertically due to tectonic plate movements, it can displace a large volume of water, generating powerful waves.

2) *Undersea volcanic eruptions*: volcanic eruptions occurring beneath the or near the ocean can also displace water, leading to the formation of tsunamis.

3) *Submarine landslides*: landslides can also occur on the seafloor just like on land. When an underwater landslide occurs following an earthquake or volcanic activity, a large mass of sand, mud and gravel can move down the slope and displaces huge volumes of water, generating a tsunami.

4) *Meteorite impacts* or *extra-terrestrial collision*: in rare cases, the impact of a large meteorite or asteroid in the ocean can generate tsunamis due to the immense energy and disturbance it creates.

2.5.3. *Case studies of tsunamis*

Several significant tsunamis have occurred throughout history, causing widespread devastation and loss of life. Here are a few notable case studies:

– *Indian Ocean Tsunami (2004)*: on December 26, 2004, a massive undersea earthquake with a magnitude of 9.1–9.3 offs the west coast of northern Sumatra, Indonesia. The tsunami generated by the earthquake affected the coastal regions of 14 countries bordering the Indian Ocean. More than 230,000 people lost their lives, making it one of the deadliest natural disasters in recorded history. The impact reached as far as East Africa.

– *Sumatra-Andaman Islands Earthquake (2005)*: on March 28, 2005, there was a powerful earthquake with a magnitude of 8.7 offs the west coast of northern Sumatra, Indonesia. Although the earthquake did not result in a destructive tsunami, it triggered a widespread tsunami warning across the Indian Ocean. The response demonstrated the importance of effective early warning systems and international cooperation. Fortunately, the tsunami generated by this earthquake was relatively small compared to the 2004 Indian Ocean Tsunami.

– *Tohoku Earthquake and Tsunami (2011)*: on March 11, 2011, a powerful earthquake with a magnitude of 9.0 offs the coast of Japan generated a massive tsunami. The tsunami inundated the northeastern coast of Japan, causing extensive damage to infrastructure and triggering a nuclear crisis at the Fukushima Daiichi Nuclear Power Plant. Over 15,000 people lost their lives, and many more were displaced. The event highlighted the need for improved tsunami warning systems and disaster preparedness.

2.5.4. *Effects of tsunamis*

The effects of a tsunami, which is a series of ocean waves with very long wavelengths (typically caused by undersea earthquakes or volcanic eruptions), can be catastrophic, affecting both the natural environment and human settlements. Here are the key effects of Tsunamis especially on vulnerable coastal communities:

1) *Loss of life and injury*: tsunamis can cause significant loss of life, property and injuries, particularly in densely populated coastal areas where people may not have sufficient time to evacuate.

2) *Destruction of infrastructure*: tsunami waves can demolish buildings, homes, utilities, roads, bridges, agricultural land and other critical infrastructure, leading to widespread destruction and significant economic losses for communities affected by tsunamis.

3) *Flooding and contamination*: the inundation of coastal areas by tsunami waves can lead to widespread flooding, carrying debris, sediment and pollutants that can contaminate soil and water sources, posing health risks. In the aftermath of tsunamis, there is an increased risk of waterborne diseases, lack of access to healthcare and mental health challenges.

4) *Economic impact*: the destruction of infrastructure, loss of livelihoods and disruption of economic activities in coastal areas can have significant economic repercussions for local communities and the broader regions.

5) *Displacement of communities*: tsunamis can force people to flee their homes, leading to displacement and potentially creating humanitarian crises as communities seek shelter and support. Therefore, many individuals and families may be displaced from their homes, leading to temporary or long-term displacement and the loss of livelihoods.

6) *Environmental damage*: tsunamis can cause extensive damage to environment and coastal ecosystems, including coral reefs, mangroves and fisheries, impacting biodiversity and the livelihoods of coastal communities.

2.5.5. Mitigation and preparedness

Efforts to mitigate the impacts of tsunamis and enhance preparedness include:

1) *Development of early warning systems*: using seismic monitoring data helps to maintain an early warning system to alert coastal populations about impending Tsunami events, allowing for timely evacuation and preparation. The Indian Tsunami Early Warning Centre (ITEWC) is located in Hyderabad at the Indian National Centre for Ocean Information Sciences (INCOIS). It comprises a real-time network of seismic stations, tide gauges and a 24 × 7 operational tsunami warning center to detect tsunamigenic earthquakes, monitor tsunamis and provide timely advisories to vulnerable communities.

2) *Land use planning and zoning*: this includes developing and enforcing land use planning and zoning regulations to restrict or prohibit construction in high-risk coastal areas for development, reducing human exposure to potential tsunami impacts.

3) *Building flood barriers*: seawalls and coastal embankments are examples of structural measures that can be taken to lessen the impact of tsunamis on coastal towns and vital infrastructure.

4) *Enforcing stringent building norms*: standards for coastal construction to guarantee that infrastructure and buildings are made to withstand tsunami effects are known as "retrofitting" or "building codes". Reducing vulnerability can also be achieved by retrofitting current structures to increase their resistance to tsunamis.

5) *Nature-based approaches*: these include preserving and restoring natural coastal buffers such as mangroves, dunes and wetlands, which can

act as natural barriers and absorb some of the energy from tsunami waves. They also include maintaining and repairing coastal ecosystems, such as mangroves, coral reefs and coastal wetlands, which stabilize shorelines, absorb wave energy and provide habitats for marine life, making them natural barriers against tsunamis.

6) *Infrastructure resilience*: designing critical infrastructure, such as hospitals, power plants, and evacuation routes, to be more resilient to tsunami impacts through elevated construction, flood barriers, and strategic placement can help mitigate the impact of tsunamis.

7) *Public awareness and preparedness*: these include raising awareness and educating coastal and vulnerable communities through drills, training and awareness campaigns that can save lives and minimize the impact of tsunamis and the potential risks, evacuation procedures and preparedness measures to ensure that they are well informed and ready to respond in the event of a tsunami warning.

2.5.6. *Safety tips*

Adhering to the dos and don'ts can help individuals and communities stay safe and respond effectively in the event of a tsunami warning or actual tsunami occurrence.

2.5.6.1. *Dos*

– *Stay informed*: pay attention and stay alert to tsunami warnings and advisories from local authorities and emergency management agencies.

– *Evacuate promptly*: if a tsunami warning is issued, people in coastal areas should move to higher ground or inland as quickly as possible, following designated evacuation routes.

– *Seek higher ground*: if you are near the coast and feel an earthquake that lasts for more than 20 seconds move to an elevated location several meters above sea level immediately as it may indicate an impending tsunami.

– *Help others*: assist children, elderly individuals and people with disabilities in evacuating to safety, if possible.

– *Have an emergency kit*: prepare an emergency kit with essential supplies such as nonperishable food, water, medications, first aid supplies and important documents.

– *Listen to authorities*: follow instructions from local authorities and emergency responders regarding evacuation, sheltering and safety measures.

2.5.6.2. *Don'ts*

– *Do not wait*: avoid waiting or hesitating if a tsunami warning is issued. Immediate action to move to safety is crucial.

– *Do not return prematurely*: do not return to coastal areas until authorities declare it safe to do so, even if the initial wave has passed, as subsequent waves may follow.

– *Do not ignore warnings*: take all tsunami warnings seriously and do not disregard official advisories as tsunamis can be extremely dangerous and unpredictable.

– *Do not rely on tsunami myths*: avoid relying on misconceptions or myths about tsunamis, and instead, follow the guidance provided by emergency management professionals.

– *Do not approach the coast*: refrain from going to the coast to observe the oncoming tsunami, as this puts you at risk of being caught in the waves.

By following these dos and don'ts, individuals can increase their chances of staying safe during a tsunami and minimizing the risks associated with these powerful natural events.

2.6. Volcanic eruptions

2.6.1. *Meaning and concept*

Volcanic eruptions occur when magma, gas and rock from within the Earth erupt onto the surface from the volcanic vent. Volcanic eruptions are caused by the release of pressure from molten rock (magma) beneath the Earth's surface, often triggered by tectonic activity or the accumulation of gases within the magma chamber. It is an endogenic process that usually occurs at plate margins caused by the release of magma from beneath the Earth's crust. They can be effusive type of eruptions characterized by the relatively gentle outpouring of lava, which can lead to the formation of lava flows and lava domes, or they can be explosive types of eruptions involving violent ejection of volcanic material, such as ash, rock fragments and gases, often leading to the formation of ash plumes, pyroclastic flows and volcanic bombs.

What is a volcano? A volcano is vent or opening where molten rock material (Ash, lava, steam and other gases) come out through the Earth's surface. It is a rupture in the Earth's crust that allows hot magma, volcanic ash and gases to escape from a magma chamber beneath the surface. Volcanoes are formed by the accumulation of erupted materials, such as lava flows, volcanic ash and tephra, which build up to create a volcanic cone. This is a process through which volcanic gases and ash are expelled from the Earth's crust onto the surface, often resulting in the formation of volcanic landforms such as lava plateaus. This phenomenon is known as volcanism or volcanicity.

Figure 2.9. *Volcanic eruptions. For a color version of this figure, see www.iste.co.uk/akbar/disastermanagement.zip*

2.6.2. *Classification of volcanoes*

2.6.2.1. *Classification based on periodicity of eruption*

1) *Active volcanos*: an active volcano is one that has recently erupted and there is a possibility that it may erupt soon, that is, Barren Islands in India, and Etna and Stromboli in the Mediterranean Sea.

2) *Dormant volcano*: a dormant volcano is one that has not erupted in a long time but there is a possibility it can erupt in the future, for example, Vesuvius volcano in Italy and Cotopaxi in South America are the best examples in this category.

3) *Extinct volcano*: an extinct volcano is one that has erupted thousands of years ago and there is no possibility of erupting again as it is empty. These are also known as dead volcanoes as they have stopped erupting but retain the features of volcanoes, for example, Kilimanjaro (Africa).

2.6.2.2. *Classification based on the mode of eruption*

1) *Central eruption type*: this is when the molten material comes out through a central pipe and form conical shape or structure. These types of eruptions are rapid and violent and may cause huge destruction and devastation. It can be divided into five sub-types such as:

a) *Strombolian type*: this type of eruption is characterized by frequent, moderately explosive bursts of lava. The lava is more viscous than in Hawaiian eruptions, leading to the ejection of incandescent cinder, lapilli and lava bombs.

b) *Hawaiian type*: these eruptions are characterized by the outpouring of highly fluid lava. The lava flows easily, creating lava fountains and flows that can travel great distances.

c) *Volcanian type*: volcanian eruptions are short, violent bursts of viscous lava. They often result in the ejection of large amounts of ash and volcanic gases.

d) *Vesuvius type*: this is considered to be one of the most dangerous volcanoes in the world for its potential to produce explosive eruptions, pyroclastic flows and lahars (mudflows). These hazards pose a significant risk to the densely populated surrounding areas.

e) *Peleean type*: these are extremely explosive eruptions that eject vast amounts of ash and pumice into the atmosphere. Peleean eruptions can produce towering columns of ash and gas reaching into the stratosphere.

2) *Fissure eruption type*: in this type of eruption, the magma comes out slowly through long fissures or cracks and then molten lava spreads over the ground surface with broad base, for example, lake fissure eruption in Iceland.

2.6.2.3. *Classification based on nature of eruption and form/shape developed*

There are several types of volcanoes, such as shield volcanoes, composite volcanoes and cinder cone volcanoes, each with distinct characteristics and eruption styles.

1) *Cinder cones*: cinder cones are the circular or oval cones made up of small fragments of lava from a single vent that has been blow up.

2) *Composite volcanoes*: composite volcanoes are steep-sided volcanoes constructed of multiple layers of volcanic materials, typically consisting of ash, rock fragments and high-viscosity lava.

3) *Shield volcanoes*: shield volcanoes are volcanoes shaped like a bowl or shield in the middle with long gentle slopes made by basaltic lava flows.

4) *Caldera*: this can result from a massive volcanic eruption or the collapse of the surface due to the emptying of the underlying magma chamber. Calderas are significant geological features and can have various forms and origins.

2.6.3. *Causes of volcanic eruptions*

Primary causes of volcanic eruptions include the following:

1) *Tectonic plate movement*: the movement of tectonic plates can lead to the build-up of pressure in the Earth's crust, causing magma to rise and eventually erupt through a volcano.

2) *Subduction zones*: when one tectonic plate moves beneath another plate, the process is called subduction. It can cause the release of pressure and the melting of rock, leading to the formation of magma and subsequent volcanic activity.

3) *Hotspots*: hotspots are areas where plumes of hot magma rise from deep within the Earth's mantle, creating volcanic activity. The Hawaiian Islands are a notable result of hotspot volcanism.

4) *Magma chamber pressure*: the pressure from the accumulation of magma in a volcano's magma chamber can lead to an eruption when it overcomes the resistance of the Earth's crust.

5) *Volcanic unrest*: changes in gas emissions, ground deformation and seismic activity can signal increased volcanic activity, leading to an eruption.

6) *Human-induced activity*: human activities, such as mining, drilling, or reservoir-induced seismicity, can lead to volcanic eruptions in some cases.

2.6.4. *Vulnerable areas*

India is not currently a hotspot for active volcanic eruptions. These areas serve as evidence of significant volcanic activity in the country's geological history. The Western Ghats mountain range in western India contains several volcanic features, including the UNESCO World Heritage Site of the Western Ghats. The region has experienced volcanic activity in the past, leading to the formation of basaltic lava plateaus and rock formations. The only active volcano in South Asia is Barren Island, which is situated in the Andaman Sea. It is a part of the Andaman and Nicobar Islands territory of India. Barren Island has experienced several eruptions in recent decades, making it the only consistently active volcano in India. In addition, Deccan Traps in west-central India are one of the largest volcanic provinces in the world. This vast region is made up of multiple layers of solidified basaltic lava flows, indicating extensive volcanic activity in the region around 60 million years ago.

2.6.5. *Effects of volcanic eruptions*

Volcanic eruptions and volcanoes are natural phenomena that can have significant impacts on the environment, climate and human populations. The hazardous effects of volcanoes are as follows:

1) *Impact on atmosphere and environment*: volcanic eruptions can release large amounts of ash and gases into the atmosphere. Volcanic eruptions release large quantities of gases, including sulphur dioxide, carbon dioxide, hydrogen sulfide and other toxic fumes into the atmosphere affecting air quality, climate and the surrounding ecosystems. These gases can contribute to air pollution, and acid rain that poses health hazards that contribute to climate and environmental effects. In addition, these events

destroy vegetation and can also cause wildfires and mass extinction of a few plants and animal species leading to overall ecological mass destruction.

2) *Lahars*: a lahar is a fast-moving mixture of volcanic debris, ash and water that flows down the slopes of a volcano, often following river valleys. Lahars can cause extensive damage to infrastructure and settlements in their path.

3) *Crop and livestock losses*: vegetation covered with ash fall is not suited for grazing. Crop damage can also affect agriculture and result in financial losses for local populations.

4) *Infrastructure damage*: volcanic eruptions can lead to the loss of life displacement of communities residing near active volcanoes, destruction infrastructure, including roads, bridges, railways and buildings, leading to disruptions in transportation and communication networks.

5) *Impact on water bodies*: volcanic eruptions that occur close to or near to water bodies lead to the degradation and contamination of watersheds, catchment areas and natural drainage systems.

6) *Impact on economy*: volcanic eruptions can have significant economic repercussions, including damage to agriculture, tourism and industry in affected regions.

2.6.5.1. *Positive effects of volcanic eruptions*

Volcanic eruptions, despite being associated with destruction and hazards, can also have positive effects on the surrounding environment, geological processes and even human societies. Some of the positive effects of volcanic eruptions include the following:

1) *Fertile soil formation*: over time, volcanic eruptions contribute to the creation of fertile soil through the deposition of volcanic ash and other minerals. These nutrient-rich soils are beneficial for agriculture and can support the growth of crops and provide benefits to ecosystems.

2) *Geothermal energy*: volcanic activity provides a source of geothermal energy. Geothermal reservoirs are heated by magma or hot rock beneath the Earth's surface. Geothermal energy, harnessed through hot springs, geysers and geothermal power plants, provides a sustainable and renewable energy source.

3) *Creation of new landforms*: volcanic eruptions contribute to the formation of new landforms such as islands, volcanic lakes, hills, mountains and lava plateaus. These landforms can have ecological and environmental significance, providing habitats for diverse species. In addition, topography produced by volcanic eruptions attracts tourists.

4) *Habitat creation*: volcanic islands and regions can provide unique habitats for diverse flora and fauna, contributing to high levels of biodiversity and endemism.

5) *Mineral resources*: volcanic activity can lead to the formation of valuable mineral deposits. Minerals such as gold, silver, copper and sulfur are often associated with volcanic environments and can be economically important.

6) *Tourism and recreation*: volcanic landscapes and features produced during eruptions often attract tourists and provide opportunities for recreational activities such as hiking, photography and nature exploration. This contributes to local economies, providing recreational opportunities and overall development.

2.6.6. *Mitigation measures or risk reduction measures*

Mitigation measures for volcanic eruptions are crucial for reducing the impact of these natural events on human life, infrastructure and the environment. Several strategies and preparedness measures can be employed to mitigate the risks associated with volcanic eruptions. Here are some key mitigation measures:

1) *Monitoring and early warning systems*: installation of monitoring systems and effective early warning systems to detect precursory signs of volcanic activity, such as changes in gas emissions, ground deformation and seismicity, can provide early warnings to local communities and authorities. Additionally, seismometers, sensors, smoke detectors, satellite monitoring and images are used to track volcanic activity.

2) *Volcano hazard mapping and zoning*: this involves the identification of vulnerable areas and establishing hazard zones of volcanic activity to avoid construction and other human activities within these zones can help to minimize the potential risks.

3) *Public awareness and education*: promoting awareness and education to local communities through discussions and other forms of media concerning volcanic risks, emergency procedures and evacuation routes can help them to prepare to respond effectively to volcanic emergencies. It also includes raising awareness of volcanic risks and preparedness measures within vulnerable areas.

4) *Infrastructure planning and resilience*: designing and constructing infrastructure with volcanic hazards in mind, such as incorporating ash-resistant building materials and designing drainage systems to handle potential lahars (volcanic mudflows), can enhance resilience to volcanic impacts.

5) *Emergency preparedness and response*: developing and regularly practicing emergency response plans, including evacuation procedures, sheltering and medical and emergency responses, can help minimize casualties and injuries during volcanic catastrophes. It also includes developing strategies to mitigate the impacts of ash fall on infrastructure and agriculture, and establishing clear evacuation routes and procedures for at-risk communities.

6) *Risk assessment*: conducting ongoing research and risk assessment to understand volcanic processes and evaluate volcanic hazards can contribute to more effective mitigation measures and public safety.

7) *Land use planning*: zoning regulations and land use planning around active volcanoes can help minimize the risk to human settlements and infrastructure.

2.6.7. *Safety tips*

Adhering to safety tips can significantly contribute to preparedness and enhance resilience during volcanic eruptions. It is essential to stay informed, remain vigilant and follow the guidance of authorities to mitigate the risks associated with volcanic activity.

2.6.7.1. *Dos*

– *Stay informed*: stay updated with information from local authorities, emergency services and reputable news sources regarding the status of the volcano, and stay indoors until any evacuation orders or safety guidelines are given.

– *Follow evacuation orders*: if the authorities issue an evacuation order, follow it promptly and evacuate to designated safe areas.

– *Protect yourself from ash-fall*: stay indoors to avoid exposure to volcanic ash, and cover windows and door openings inside homes, which can be harmful if inhaled. If you must go outside, wear a mask and protective clothing.

– *Prepare an emergency kit*: have an emergency kit ready with essentials such as water, nonperishable food, first aid supplies, flashlights, batteries and important documents and be ready for evacuation.

– *Stay clear of restricted areas*: respect restricted zones established by authorities and avoid entering hazardous areas around the volcano.

2.6.7.2. *Don'ts*

– *Don't ignore warnings*: ignore official warnings or evacuation orders. It is essential to take these directives seriously for your safety.

– *Don't return prematurely*: return to evacuated areas prematurely without clearance from authorities, as there may still be ongoing hazards.

– *Don't drive through ash fall*: drive through or operate machinery in heavy ash fall, as it can cause visibility issues and damage engines.

– *Don't drink contaminated water*: consume water that may be contaminated by volcanic ash. Instead, rely on stored or treated water.

– *Don't venture into volcanic craters or vents*: enter volcanic craters or attempt to explore volcanic vents, as these areas can be extremely hazardous and unpredictable.

– *Don't underestimate secondary hazards*: underestimate secondary hazards such as lahars, pyroclastic flows and mudflows, as they can occur even after the initial eruption.

2.7. Wildfires

2.7.1. *Meaning and concept*

Wildfires are uncontrolled and unplanned fires in forest areas that wipe out large fields and land areas, destroying almost every organic matter in it. Wildfires are the most common type of hazards in forests. There are times when these fires linger for days or weeks. They can be termed as forest fires, grass fires, peat fires, bush fires and hill fires, depending on the type of vegetation being burnt. These fires occur in areas such as forests, grasslands or prairies and can be ignited by natural causes like lightning or by human activities such as campfires, discarded cigarettes or arson. Abundant dry vegetation, such as grass, shrubs and trees, acts as fuel for wildfires. Accumulated dead vegetation can significantly contribute to the spread and intensity of wildfires, destroying forest lands and can be dangerous to the people who live in or near the forests as well as wildlife. These hazards are typically elevated during dry seasons when vegetation is parched and more prone to ignition. During the summer when there is no rain for months (usually the summer season is from April to June), the conditions become favorable for the fires to take place. This is mainly because the temperature during that period goes beyond 40°C, making everything so dry the forests become littered with dry sensitive leaves and twigs which could turn into flames ignited by the slightest spark, further, the wind speed during this period is also very high, which intensifies and spread fires rapidly. Extended periods of hot, dry weather increase the risk of wildfires. High temperatures and low humidity create favorable conditions for the ignition and spread of fires. Understanding the timing and seasonality of wildfire hazards is crucial for proactive wildfire prevention, preparedness and effective emergency response planning.

Figure 2.10. *Wildfires. For a color version of this figure, see www.iste.co.uk/akbar/disastermanagement.zip*

2.7.2. *Types of wildfires*

There are various types of wildfires based on their characteristics and the environment in which they occur. Some common types include the following:

1) *Surface fires*: these fires typically burn the surface of the ground. They burn dry grass, broken twigs and branches dry leaves, and other low-lying vegetation. They are the most common type of wildfires, and generally move at a moderate speed with flames almost one to two meters high.

2) *Crown fires*: these are the fires in which the crown of the trees and shrubs burn. These fires spread from treetop to treetop and often spread very quickly with wind and heat. These are the most dangerous wild or forest fires and can burn trees up to their entire length to the top.

3) *Ground fires*: these fires occur underground in the organic layers of soil, often ignited by lightning strikes. They can smoulder for extended periods and are difficult to detect and extinguish.

4) *Urban interface fires*: these fires occur in or near urban areas where structures and communities interface with wildland areas. They pose significant risks to both human populations and infrastructure.

2.7.3. *Causes of wildfires*

Some common causes including both natural and man-induced causes are discussed as follows.

2.7.3.1. *Natural causes*

– *Lightning*: when lightening occurs, it strikes tree and power lines producing a spark causing wildfires. Lightning strikes are a frequent natural cause of wildfires, especially in dry regions with thunderstorms.

– *Spontaneous combustion*: the decomposition of organic matter, such as peat, can generate heat and lead to spontaneous ignition under certain conditions.

– *Wind conditions*: strong winds can rapidly spread wildfires, making them more challenging to control and extinguish.

– *Topography*: steep slopes and rugged terrain can influence the speed and direction of wildfire spread. Fires can travel uphill more quickly due to preheating and the availability of oxygen.

– *Dry weather*: prolonged periods of dryness and hot weather conditions offer favorable circumstances for fires to start quickly. Low-humidity levels can further exacerbate the dryness of vegetation, making it more susceptible to catching fire.

– *Volcanic eruptions*: hot burning lava from volcanoes cause wildfires.

– *Other natural causes*: the forest floor can also catch fire naturally due rolling of bamboo, dry branches and friction of stones producing sparks that can turn into a big disaster.

2.7.3.2. *Human-induced causes*

Fire is caused when the source of fire or ignition comes into contact with flammable materials. Human activities, such as using campfires, discarding cigarettes, sparking objects or deliberate arson, can also ignite wildfires. Additionally, activities such as land clearing or burning of agricultural debris and yard waste without proper precautions can lead to uncontrolled fires.

– *Unattended campfires*: people of all ages spend time in the woods to enjoy nature, and camping is quite popular in many areas. Fire is sometimes needed for several reasons. Unattended or improperly extinguished campfires have the potential to ignite nearby vegetation.

– *Arson*: this includes deliberate setting of fires by individuals for various reasons, including vandalism, insurance or malicious intent.

– *Equipment failure/accidents*: car crashes or blasting, use of gas and hot air balloons, plane crashes, engine sparks and operation of machinery, such as chainsaws or vehicles, in dry vegetation, can create sparks and ignite fires.

– *Smoking*: smoking can cause wildfires if carelessly discarded or if a lit cigarette comes into contact with dry vegetation. The risk is especially high in areas prone to wildfires or during dry, windy conditions. Improper disposal and carelessly discarded cigarette butts can smoulder and spark, leading to fires in dry environments.

2.7.4. *Forest fire prone regions of India*

India has several regions prone to wildfires or forest fires, with varying levels of susceptibility and vulnerability to fire depending on the type of vegetation and climate: Himachal Pradesh, Uttarakhand, M.P. The pine forests in the Grewal and Kumaon hills are the most susceptible to forest fires, more than any other kind of forest. For instance, coniferous forests are 8% prone to frequent fires and 40% to occasional fires; moist deciduous forests are 15% prone to frequent fires and 60% to occasional fires; dry deciduous forests are 5% prone to frequent fires and 35% to occasional fires; wet/semi-evergreen forests are 9% are prone to frequent fires and 40% to occasional fires; and the northeastern region is 50% prone to frequent fires and 45% to occasional fires. Various regions of the country have different normal and peak fire seasons, which typically vary from January to June. In the plains of northern and central India, most forest fires occur between February and June. In the hills of northern India, the fire season starts later and most fires are reported between April and June. The southern region of the nation experiences the fire season from January to May. May and June are fire seasons in the Himalayan region.

2.7.5. *Impact of wildfires*

The impact of wildfires or forest fires is disastrous and devastating and has long-term effects destroying large areas and everything in its way.

1) *Impact on environmental/atmosphere*: wildfires release large amounts of smoke (toxic fumes) and particulate matter into the atmosphere, which can degrade air quality. Also some ecosystems may struggle to recover after severe fires, impacting their ability to regenerate and support wildlife. Wildfires release large amounts of carbon dioxide into the atmosphere, contributing to greenhouse gas emissions and overall climate change.

2) *Soil erosion*: forest soils are rich in nutrients and composed of organic matter that support life forms. Wildfires raise the temperature of the soil and this potentially wipes away almost all organic value and fertility of the soil. The loss of vegetation due to wildfires can lead to increased soil erosion, especially in steep terrains.

3) *Loss of biodiversity*: wildfires can destroy ecosystems and habitats of living organisms and can lead to the loss of various plant and animal species, disrupting ecological balance and biodiversity.

4) *Economic impact*: wildfires can damage property such as homes, infrastructure, agricultural lands and timber resources resulting in significant economic losses. Also, the cost of firefighting efforts, including personnel, equipment and resources, can be substantial.

5) *Impact on human health*: smoke and ash from wildfires can exacerbate respiratory problems or breathing issues and pose health risks to people, especially those with preexisting medical conditions.

6) *Community and social impact*: wildfires can have a lasting emotional and psychological impact on individuals and communities, especially those who have lost homes or loved ones. The social life of communities can be disrupted and recovery may take years.

7) *Water quality*: wildfires can degrade water quality by increasing sedimentation and contaminating water sources like rivers, streams, ponds and nearby lakes with ash and debris.

2.7.6. *Mitigation measures*

Efforts to mitigate the risk and impacts of wildfires include the following:

– *Early monitoring and detection*: a proper monitoring and detection system should be operated to give warning and information to the communities. This would help in avoiding panic and ensuring that people take necessary safety actions in case of emergency situations. Apart from this, danger rating and real-time monitoring have to be designed and installed. Instead of fire watchers, satellite imagery, CCTV cameras and drones can be used to monitor wildfires, and sensors and smoke detectors can be used for detecting smoke. In addition, investment is made in early detection systems such as aerial surveillance and remote sensing technologies to detect wildfires at their inception for quick response to wildfires.

– *Community planning and preparedness*: building codes and zoning regulations are developed and enforced, taking wildfire risks into account. Communities should be educated about wildfire preparedness, evacuation plans and the importance of creating fire-resistant landscapes.

– *Public awareness and education*: this includes educating the public about the causes and effects of wildfires on the environment, fire safety, evacuation procedures and the importance of fire-resistant building materials. Establishing training centers and arranging firefighting drills frequently to provide practical training for efficient application of the equipment are also important.

– *Firefighting resources*: this includes maintaining well-equipped and trained firefighting personnel. It also includes use of aerial firefighting resources, such as helicopters and air tankers or aerial vehicles, to combat wildfires in the affected area.

– *Use of artificial barriers and firebreaks*: this includes creating barriers to slow or stop the progress of a wildfire. Forests can be divided into compartments bordered by natural barriers such as streams, roads, fire lines and corridors. If the blaze has burnt the affected compartments, it vanishes out and the neighboring compartments can be saved.

– *Legislation and policy*: this includes enacting and enforcing policies related to fire prevention, land use planning and forest management. It also includes supporting research and development of new technologies and strategies for wildfire mitigation.

– *Effective communication and coordination*: this includes setting up of better communication and coordination systems, and fostering partnerships between government agencies such as national remote sensing agencies, firefighting organizations, community groups, forest departments, environment ministries and private land-owners to coordinate wildfire mitigation efforts and for a prompt and coherent response.

2.8. Droughts

2.8.1. *Meaning and nature*

According to the McGraw Hill (2005) *Concise Encyclopedia of Science and Technology*, droughts are a general term implying a deficiency of precipitation (rain or snow) of sufficient magnitude to interfere with some phase of the local economy.

Widespread or prolonged droughts are one of the common and dreadful natural disasters causing crop failure due to lack of rain. Sometimes, this can lead to famine to make conditions worse. This simply refers to a prolonged period in which a region has abnormally low precipitation leading to water shortages and parched soil for an unusually long period from its normal pattern, which can have significant impacts on the environment, agriculture and human populations. During this period, food and water supplies can run low, and other conditions, such as famine, can also result.

Figure 2.11. *Parched earth due to a drought. For a color version of this figure, see www.iste.co.uk/akbar/disastermanagement.zip*

Droughts are destructive, particularly in places where people rely on agriculture for living and can last for several years. Droughts can have unpredictable beginnings and ends, and extended ones can cause disruptions to the availability of food and water and can even force people to migrate against their will. Drought conditions may also lead to famines characterized by a widespread lack of food, livestock and agricultural food stuff required for basic nutrition and survival in a region. In addition, its effects may be exacerbated by social factors, such as war and conflicts. A drought can cover areas from a few hundred square kilometers to vast regions of hundreds of thousands, or in extreme cases, millions of square kilometers. When they occur over comparatively smaller areas of a few hundred square kilometers or so, the severity and duration are generally less than those expected in case of droughts covering large regions of thousands of square kilometers. We could also say that the more severe a drought, the larger its area coverage. The beginning and end of a drought, especially the beginning, are not sharp cut-off points in time, although a drought could sometimes end rather sharply with a wet spell of heavy widespread rain. Therefore a drought is a phenomenon with a slow onset. Mostly, it is a creeping phenomenon that builds up slowly and steadily. Due in part to its self-perpetuating nature, droughts occur frequently. This is because there is more heat on the ground surface and fine dust in the sky once the drought has begun and the land surface is bare of vegetation due to dryness and overgrazing.

2.8.2. *Types of droughts*

There are different types of droughts according to the extent of water scarcity, and meteorological, hydrological and agricultural are the most common types. Let us discuss them in detail:

1) *Meteorological drought*: this is explained as the condition when actual rainfall over a region is less than 75% of the normal value or situation when there is a reduction in rainfall for a specified period (day, month, season or year) below a specified amount, often expressed as a percentage of the long-term average for the given time frame.

2) *Hydrological drought*: this is explained as the condition when water scarcity leads to a situation marked with depletion of surface water causing low stream flow and the drying of water bodies such as lakes, reservoirs and rivers. It may also result in recession of mountain springs and glaciers due to insufficient regeneration of seasonal snow cover on account of lack of precipitation (rain or snow) or simply involves a reduction in water resources (stream flows, lake levels, groundwater and underground acquirers) below a specified level for a given period.

3) *Agricultural drought*: when water scarcity becomes so acute that the soil moisture is not adequate to support the healthy growth of crops. The water table recedes (i.e. goes deeper) and irrigation needs cannot be fulfilled by groundwater also. Actually, it is the impact of meteorological and/or hydrological droughts on crop yields during the growth cycle. This type of drought occurs when the soil moisture is not sufficient to support the production of crops, which causes extreme crop stress and wilting during crop development or growth period and can lead to starvation.

2.8.2.1. *Based on the climate of the region and duration*

1) *Long-term drought*: long-term droughts persist for several years and can have widespread and severe impacts on water resources, agriculture, ecosystems and overall societal well-being. It generally refers to a prolonged and severe lack of precipitation that persists for an extended period. This season is the driest of all, and agriculture is impossible without constant irrigation.

2) *Short-term drought*: this type of drought lasts for a few weeks to a few months. Short-term droughts can also impact agriculture, water supply and surface water levels.

3) *Seasonal drought*: this occurs in climates that have distinct and well-defined seasons. Seasonal droughts occur during specific seasons and can affect agricultural activities, particularly in regions with distinct wet and dry seasons. Here agricultural practices have to be adopted so that they take advantage of the rainy season.

4) *Contingent or unpredictable drought*: this involves an abnormal failure of rains. It can occur anywhere even in the regions which receive heavy rains, for example, Cherrapunji.

2.8.3. *Causes of droughts*

The main causes of droughts can be attributed to a variety of factors, including natural causes and human-induced causes:

1) *Natural causes*: natural fluctuations like climate change and weather patterns, such as El Niño and La Niña, can lead to periods of reduced precipitation, contributing to drought conditions in some regions. Extended periods of below-average rainfall or snowfall can lead to water shortages. In addition, persistent pressure systems can block the formation of rain clouds, contributing to prolonged dry spells, and also high rise in temperature can exacerbate evaporation rates, leading to increased water loss from soils and vegetation and reducing water availability. Mountains can block the passage of rain-producing weather systems, leading to decreased precipitation on the leeward side, causing drought conditions. Changes in climate patterns, such as increasing temperatures and altered precipitation regimes, can contribute to prolonged droughts.

2) *Human-induced causes*: human activities, such as the burning of fossil fuels and deforestation, global warming contributes to climate change, which can alter precipitation patterns and exacerbate drought conditions. Excessive use of water for agriculture, industry and domestic purposes can deplete water sources and contribute to drought conditions. Intensive farming practices, including irrigation without proper water management, can deplete soil moisture, leading to agricultural droughts. In addition, overutilization of underground water and poor water management practices can also lead to droughts. The removal of trees and vegetation can disrupt local water cycles, leading to reduced precipitation and increased susceptibility to droughts.

2.8.4. *Drought-prone regions of India*

India's varied climatic and geographical characteristics result in recurring drought conditions in several of its regions. Droughts are frequently linked to the little erratic rainfall, which causes water scarcity, problems for agriculture and other socioeconomic issues. Given the large size of the country and the very high time and space variability of the southwest monsoon rainfall, it is not surprising that no part of India can be regarded as free from the likelihood of a drought. This is why there are parts of the country suffering from droughts even in the best of the monsoon years (e.g. Madhya Pradesh in 1988) or sometimes droughts occur in the usually very rainy northeastern states of the country in 2005. India's drought-prone areas include some of the following:

– *Rajasthan*: the dry and semi-arid climate of the Rajasthan state, located in northwest India, is its defining feature. Droughts are very common in the Rajasthan regions of the Thar Desert. In this area, desertification and water scarcity are major problems.

– *Gujarat*: during the weak monsoon seasons, some parts of Gujarat, particularly the Kutch region, endure semi-arid conditions and water shortages. In Gujarat, droughts can affect the water supply for both rural and urban areas, as well as agricultural productivity.

– *Maharashtra*: parts of Maharashtra, including the Vidarbha and Marathwada regions, are prone to recurring droughts. Agriculture-dependent communities in these areas often face water stress, crop failures and socioeconomic challenges during periods of inadequate rainfall.

– *Karnataka*: several districts in Karnataka, particularly in the northern and eastern regions, are prone to drought conditions. The scarcity of water affects agriculture and can lead to distress in rural communities.

– *Andhra Pradesh and Telangana*: certain districts in both Andhra Pradesh and Telangana, especially those in the Rayalaseema region, face drought conditions during deficient monsoon years. Water scarcity impacts agricultural practices and can lead to farmer distress.

– *Tamil Nadu*: some parts of Tamil Nadu, particularly in the southern and western regions, experience water scarcity and drought conditions. Chennai, the capital city, has faced water shortages, emphasizing the need for effective water management strategies.

– *Uttar Pradesh*: parts of Uttar Pradesh, particularly the Bundelkhand region, are susceptible to drought conditions. Water stress and inadequate rainfall can impact agricultural activities and rural livelihoods.

2.8.5. *Impact of droughts*

Droughts can have wide-ranging impacts on the environment, economy and society, including:

1) *Agricultural impact*: this includes reduced crop yields and livestock productivity due to water scarcity and soil moisture deficits. Crop failures lead to food shortages, higher food prices, potential food insecurity and loss of agricultural income and livelihoods for farmers and agricultural workers. Deterioration of wetlands, rivers and other aquatic ecosystems due to reduced water availability lead to habitat loss for aquatic species. Degradation of terrestrial habitats, such as grasslands and forests, is a result of increased stress on plant species and loss of niche habitats. Reduction in ecosystem services such as pollination, water purification, and soil stabilization impact overall ecosystem resilience.

2) *Environmental impact*: stress on natural ecosystems, including forests, wetlands and rivers, leads to reduced biodiversity and habitat degradation. Increased risk of wildfires due to dry conditions poses a threat to plant and animal species and causes environmental damage.

3) *Water supply and quality*: depletion of surface water and groundwater resources affects drinking water supplies, irrigation and industrial water usage. Increased salinity in water bodies due to reduced inflows impacts freshwater ecosystems and water quality.

4) *Economic impact*: reduced hydropower generation capacity leads to energy shortages and increased reliance on alternative energy sources. There are losses in tourism revenue in areas affected by droughts, particularly those reliant on natural attractions such as lakes, rivers and scenic landscapes. Sectors reliant on water are also impacted, such as agriculture and tourism. Changes in biodiversity and reduced vegetation cover and productivity lead to changes in plant species composition and diversity. Lastly, the impact is the disruption of ecological balance, with potential shifts in animal populations and changes in predator–prey dynamics.

5) *Wildlife and habitat degradation (health and social impact)*: this includes an increased risk of waterborne diseases and inadequate sanitation due to limited access to clean water. It also causes social and psychological stress on communities, particularly in rural areas heavily reliant on agriculture for livelihood. In addition, this causes social disruption in the form of resource-related conflicts, wars, suicide and migration.

6) *Secondary effects of droughts*: primary effects of droughts mainly result from lack of water. However, the secondary effects of droughts follow and result from the primary effects. In such circumstances, people begin to migrate in search of better grazing lands for their herds or to the cities to seek alternate sources of income. If the dwindling supplies of food are not replaced, famines can occur, further accelerating the migration. The migration may in itself contribute to the spreading of the scope of the disaster. Long-term droughts result in permanent changes in settlement patterns and in social and living patterns. Among the ecological changes, the desertification cycle is of utmost concern. The whole process gradually spreads, bringing more and more land under desert conditions.

Primary effects of droughts	**Secondary effects of droughts (short and long term)**
Loss of food crops	Famine and food insecurity
Loss of agricultural production	Spread of diseases and more deaths
Effects on nutrition	General malnutrition and mortality
Loss of livestock and other animals	Increased migrations
Water scarcity/water quality	Reduction in water levels, increased desertification
Loss of industrial production	Social disruption
Loss of hydroelectric power/energy	Changes in settlement pattern ecological imbalance
Wildfires	Soil erosion and probability of ignition – drying up of soil and plants
Reduces stream flow/water supply	Inadequate water supply, economic impact
Health issues	Increased incidence of illness or disease, higher mortality rates and adverse mental health outcomes

Table 2.3. *Effects of droughts*

2.8.6. *Drought management*

In India, the Indian Meteorological Department (IMD), National Remote Sensing Agency (NRSA), Central Water Commission (CWC) and the Ministry of Agriculture are some of the organizations responsible for drought monitoring. Managing droughts requires a combination of short-term and long-term strategies to mitigate its impacts on water resources, agriculture, public health and the environment. Through the application of appropriate drought-management measures and strategies, we can reduce various impacts associated with droughts and enhance preparedness. A drought, a natural phenomenon characterized by prolonged periods of abnormally low precipitation, poses significant challenges to ecosystems, societies and economies worldwide. As drought events become more frequent and severe due to climate change and increasing water demands, effective drought-management strategies are essential to mitigate its impacts and build resilience. There are various approaches to drought management, including water conservation, sustainable agriculture, policy interventions and community engagement.

Water conservation is a cornerstone of drought management efforts. Implementing measures to reduce water consumption, such as promoting water-efficient technologies, fixing leaky infrastructure and implementing water-saving practices in households and industries can help alleviate pressure on dwindling water resources during dry periods. Public awareness campaigns and educational programs are crucial for encouraging water conservation behaviors and fostering a culture of responsible water use among individuals and communities. We need to educate the public about water conservation practices, the impacts of droughts and the importance of efficient water use, and promote community engagement in water management and encourage responsible water consumption behaviors. Implementing measures to conserve water in households, agriculture and industries to reduce water demand, water recycling and reuse systems to maximize available water resources. Through rainwater harvesting, we can store water from rain and then we use it when we need it in dry conditions. In addition, judicious use of surface and sub-surface water recharges and improved storage facilities. These are the best practices that can be followed during drought conditions.

Sustainable agriculture practices play a vital role in drought management by minimizing water usage and enhancing soil moisture retention. Techniques such as drip irrigation, mulching, crop rotation and the use of drought-resistant crop varieties can improve agricultural water efficiency and resilience to drought conditions. Furthermore, promoting soil conservation practices like no-till farming and agroforestry can enhance soil health and water infiltration, reducing the impacts of drought on crop yields. Policy interventions at the local, national and international levels are instrumental in drought management. Governments can develop and enforce regulations for sustainable water management, including water allocation schemes, groundwater management plans and drought contingency plans. Incentive programs and subsidies can encourage the adoption of water-saving technologies and practices across various sectors. Additionally, integrating climate adaptation measures into land-use planning, infrastructure development and natural resource management can enhance resilience to droughts and other climate-related risks, implement water-use restrictions and allocation mechanisms during drought periods to ensure equitable distribution and prioritization of critical water needs, and develop and enforce regulations for sustainable water use in agriculture, industry and urban areas.

Community engagement and stakeholder collaboration are essential components of effective drought management. Engaging local communities, farmers, water users and other stakeholders in decision-making processes foster ownership and cooperation in implementing drought preparedness and response measures. Community-based initiatives, such as water user associations, drought monitoring networks and early warning systems, can enhance the local capacity to cope with drought impacts and facilitate timely interventions. Furthermore, investing in drought monitoring, early warning and risk assessment systems are critical for proactive drought management. Leveraging advances in technology, such as remote sensing, weather forecasting and data analytics, enables timely detection of drought onset, monitoring of drought severity and prediction of potential impacts. This includes creating drought monitoring and forecasting systems that use meteorological data and remote sensing to provide early warning of potential droughts, and establishing drought response contingency plans for farmers to manage water resources during periods of scarcity in a timely and effective manner. Early warning systems provide decision-makers with actionable information to implement targeted interventions, including water rationing, emergency relief measures and drought assistance programs, integrating drought management into broader climate change adaptation strategies to address long-term shifts in precipitation patterns.

By implementing a combination of short-term emergency measures and long-term sustainable strategies, communities and governments can effectively manage the impacts of droughts and build resilience to future water scarcity.

2.9. Snow avalanches

2.9.1. *Meaning and concept*

A snow avalanche is a rapid downhill movement of a huge mass of snow, ice and debris grinding everything in its path and rushing down a steep slope of a mountain. The characteristics of the terrain, such as slope angle, aspect (the direction the slope faces) and vegetation cover, significantly influence avalanche risk. Steep slopes, convex terrain and areas with multiple connected slopes increase the likelihood of avalanches. They are typically triggered by a variety of factors, such as the accumulation of snow, rapid changes in temperature, wind patterns or human activities such as skiing, snowmobiling or explosives used in avalanche control. When a critical amount of stress is applied to a snowpack, it can fail and release, resulting in an avalanche. These events can pose significant risks to human life, infrastructure and the natural environment, particularly in mountainous regions and areas with heavy snowfall. An avalanche comprises three zones, that is, the starting zone, avalanche track and runout zone. The starting zone is the point where unstable snow can fracture from the surrounding snow cover and begin to slide down the avalanche tract, it is the path followed by a particular avalanche while sliding downwards. The runout zone at the end is where the huge mass of snow finally come to stop. Snowfall can come loose in large pieces or form a wall that sweeps victims down the mountain. An avalanche can frequently be generated by loud noise vibrations, such as those produced by artillery firing, thunder or rock blasting.

Figure 2.12. *The three zones of snow of an avalanche. For a color version of this figure, see www.iste.co.uk/akbar/disastermanagement.zip*

An avalanche is a massive mass of snow or ice that falls from a slope with considerable force and speed, causing extensive damage at a large scale as it approaches and in the valley below. The downward motion may be in the form of gliding or sliding along the slope like a rock fall, flowing along the ground like fluid or whirling through the air like a whirlwind.

2.9.2. *Types of snow avalanches*

There are several types of snow avalanches, each with distinct characteristics and behaviors, such as loose snow avalanches, slab avalanches and wet snow avalanches. The main types include:

1) *Slab avalanches*: these are the most common and dangerous kinds of avalanches. A substantial block of ice that slides downhill or is separated from its surroundings by cracks causes the packed snow to become loose and flow downhill. Human activity, temperature variations or the continuous weight of fresh snow can also cause snow avalanches.

2) *Loose snow avalanches*: also known as sluffs, these are composed of unconsolidated snow that is made up of surface or near the surface snow and is not well bounded. They usually begin at one location and spread out as they descend. They are common on steep slopes and are seen after a fresh

snowfall or can be disturbed by the movement of people on the surface of snow.

3) *Icefall avalanches*: icefall avalanches occur when ice breaks from a glacier or icefall and cascades down the mountain equivalent to waterfall. They are often triggered by changes in temperature, melting or the movement of the glacier.

4) *Powder snow avalanches*: powder snow avalanches involve very fine, dry snow that flows like a fluid down the slope. They are commonly triggered by additional loading from fresh snow or by a person's weight.

5) *Wet snow avalanches*: these avalanches are triggered by warm temperatures, rain or rapid melting of the snowpack and occur when the snowpack becomes saturated with water, losing cohesion and flowing downhill, usually due to increased temperatures or rain causing water to percolate through the snowpack and decreasing the strength of snow. Wet snow avalanches are capable of generating powerful destructive force due to their large mass and density. Wet snow avalanches are quite dangerous and tend to be slower moving but can contain a large volume of snow and debris.

6) *Glide avalanches*: glide avalanches are slower-moving and are caused by the gradual gliding of a thick layer of snow over the ground. They often occur on smooth, grassy slopes and are very difficult to predict.

7) *Cornice fall avalanches*: cornice fall avalanches occur when an overhanging edge or cornice of snow collapses and triggers a cascading snow slide below. They are typically associated with ridgelines and can pose a significant hazard to climbers and backcountry travellers.

2.9.3. *Causes of snow avalanches*

Snow avalanches are typically triggered by a combination of various factors. The primary factors contributing to snow avalanches include:

1) *Heavy snowfall*: heavy snowfall, especially when it occurs rapidly, can increase the load on the snowpack, adding stress and potentially triggering avalanches. Wet, heavy snow or rain on existing snow can also contribute to instability by weakening the unstable areas of the mountain. In addition, the snowpack becomes hypersensitive as a result of the layer-by-layer accumulation of snow caused by the slow snowfall. These snow layers slide as a result of catastrophic events, causing an avalanche.

2) *Wind direction*: the direction and speed of the wind determine the patterns of the snowfall as well as snow accumulation on the mountain slopes. If the strong wind blows, then the upward direction of the winds may trigger a steep slope which is often unstable and can cause an avalanche. Changes in wind speed and wind direction due to snowstorms will also trigger snow avalanches. Wind can transport and redistribute snow, creating different layers and densities in the snowpack. Wind loading can lead to the formation of wind slabs, which are often unstable and prone to avalanching. Wind can also erode snow from leeward slopes, exposing weak layers.

3) *Higher temperatures*: weak layers within the snowpack often caused by factors like high-temperature gradients can lead to instability. Some of the top layers of snow can become weaker and slide down when it is warm and sunny for a few hours each day. In addition, rapid temperature changes, especially if they lead to freeze-thaw cycles, can affect the stability of the snowpack and can create melt-freeze crusts, which act as slippery layers and contribute to avalanches.

4) *Human activity*: human activities have contributed to the start of many avalanches in recent years. Winter sports activities such as skiing and snowmobiling on steep slopes, the movement of people and animals or loud noises – even a sharp whistle – can generate vibrations and can start the mass in motion. Mountainous regions are more vulnerable to deadly avalanches as a result of development and commercial activities. These activities usually put additional stress on snow cover and can trigger avalanches.

5) *Deforestation*: deforestation can alter the natural forest cover and disrupt the stability of snowpack in mountainous regions. Trees and plants play a crucial role in holding snow in place, preventing it from accumulating to dangerous levels. When trees are removed, the snowpack becomes more prone to instability and can lead to avalanches. In addition, it can lead to soil erosion, which results in the deposition of sediments in streams and rivers. This can affect the natural flow of water and contribute to the destabilization of snowpack in higher elevations, potentially triggering avalanches.

6) *Natural triggers*: the layers of accumulated snowpack are often triggered by natural events like earthquakes, tremors, thunder and rock falls. Earthquakes produce seismic waves, which move the ground and can induce greater avalanches. These occurrences have the potential to disrupt the snowpack and cause weak layers to fail.

2.9.4. *Snow avalanche-prone regions of India*

India is home to several regions prone to snow avalanches, especially in the northern and northeastern parts of the country where the Himalayan mountain range spans. These regions experience heavy snowfall and have steep terrain, making them vulnerable to snow avalanches. Snow avalanche is a problem of high altitude mountainous regions, which receive heavy snow. The northern regions of Jammu and Kashmir, Himachal Pradesh, the hills of Uttaranchal and the eastern region up to Sikkim are the areas of our country most vulnerable to avalanches. These areas, due to communication difficulties and remoteness from the rest of the country, are very thinly populated and under-developed and hence a lot of incidents of this nature in these areas go unnoticed and they do not receive media attention. Nevertheless, there lies a great potential for development, which needs our careful attention for mitigating such disasters. Every year, avalanches in these areas take a heavy toll on lives and cause considerable drainage to property. Some of the prominent snow avalanche-prone regions in India include the following:

1) *Jammu and Kashmir*: areas such as Gulmarg, Sonmarg and higher reaches of the Pir Panjal and Zanskar ranges are prone to snow avalanches during winter.

2) *Himachal Pradesh*: regions around Kullu, Manali, Lahaul, Spiti and Kinnaur districts often experience snow avalanches due to heavy snowfall in the higher altitudes.

3) *Uttarakhand*: high-altitude areas in Uttarakhand, including the districts of Uttarkashi, Chamoli, and Pithoragarh, are susceptible to snow avalanches, especially after heavy snowfall.

4) *Arunachal Pradesh and Sikkim*: the higher reaches of these northeastern states, including areas near the Indo-China border, are prone to snow avalanches during the winter months.

5) *Hilly areas of northeastern states*: various hilly areas in states like Meghalaya, Nagaland and Manipur are also susceptible to snow avalanches during winter.

2.9.5. *Effects of snow avalanches*

Snow avalanches can have devastating impacts on communities, infrastructure and natural habitats, causing injuries, fatalities and property damage. Snow avalanches can have significant and far-reaching impacts on both the natural environment and communities.

The consequences of a snow avalanche can vary depending on factors such as the size of the avalanche, the terrain and the presence of human settlements. Some of the potential impacts of snow avalanches are discussed as:

1) *Loss of life and injury*: avalanches can result in the loss of human lives, especially if they occur in areas with residential or recreational developments. People caught in avalanches may suffer injuries due to the force and debris carried by the moving snow.

2) *Property damage*: snow avalanches can cause extensive damage to buildings and infrastructure. This can include homes, roads, bridges and other structures. These hazards can block roads, bridges, railways and other structures disrupting transportation and making it challenging for rescue teams to reach affected areas or for individuals to evacuate.

3) *Economic impact*: the economic impact of avalanches includes the costs associated with rescue and recovery efforts, property damage and the interruption of economic activities in affected areas. Also, areas that are popular for winter tourism and recreation may experience disruptions due to avalanches, impacting local economies and loss of businesses.

4) *Environmental impact*: avalanches can alter the landscape and vegetation in their path, affecting ecosystems and wildlife habitats. They can also lead to soil erosion and changes in watercourses.

5) *Flash floods*: blockage of streams and later on impounded release of water create flash floods with disastrous effects. Falling of large volumes of debris from snow avalanches in mountain lakes can generate flash floods.

6) *Psychological impact*: the fear and trauma associated with avalanches can have a lasting psychological impact on individuals and communities affecting mental health and well-being.

2.9.6. Preparedness and mitigation measures

To reduce the impact of avalanches, communities in avalanche-prone areas implement proper land-use planning, establish warning systems and adopt safety measures, such as avalanche control programs and education initiatives for residents and visitors. Mitigation measures for snow avalanches are crucial for reducing the risk to human life, infrastructure and the environment.

Several strategies are employed to mitigate the impact of snow avalanches:

1) *Avalanche forecasting and early warning systems*: avalanches have led to the development of advanced monitoring and forecasting early warning systems. The weather is monitored using high-resolution cameras, motion sensors and modern RADAR technology, which help to predict and mitigate impacts. These systems also use technology such as weather monitoring, snowpack analysis and remote sensing to predict and warn about potential avalanche risks.

2) *Avalanche control measures*: these may include the use of explosives, artillery or other means to intentionally trigger avalanches in a controlled manner, reducing the risk of larger, uncontrolled avalanches.

3) *Structural mitigation*: this involves the construction of physical structures such as snow sheds, barriers or walls to deflect or contain avalanches and protect infrastructure and settlements.

4) *Land-use planning and zoning*: implementing zoning regulations and land use planning to restrict development in high-risk avalanche areas can help minimize exposure to avalanche hazards.

5) *Public education and awareness*: educating individuals about avalanche safety, including backcountry travel, rescue techniques and the use of avalanche safety equipment, is essential for reducing the risk to recreationists and workers in avalanche-prone areas.

6) *Vegetation and forest management*: maintaining healthy forests and vegetation can help stabilize slopes and reduce the risk of avalanches by enhancing snowpack retention and reducing snow accumulation.

2.10. Questions

1) Explain in detail flood disasters and the types of floods.

2) Discuss some preventive methods to minimize the impact of floods.

3) Describe volcanic eruptions. What are their causes and effects?

4) What are the safety measures for volcanic eruptions?

5) Define earthquakes. Classify earthquakes based on depth of focus?

6) What are the damaging effects of earthquakes and their preventive measures?

7) What is a wildfire, explain its various types.

8) What are the mitigation measures for forest fire disasters?

9) Define the following terms:

a) Seismogram

b) Subduction zone

c) Ring of Fire

d) Retrofitting

10) List the regions and seasons prone to snow avalanches in India and discuss the causes of the formation of snow avalanches.

11) What are snow avalanches, their causes and mitigation measures?

12) What are the various types of snow avalanches?

13) List the dos and don'ts that should be observed while moving in an avalanche-prone area.

14) What are landslides? Discuss some preventive methods to minimize the impact of landslides.

15) What is a drought and explain its types?

16) What are the major causes of droughts? Discuss various drought-control measures.

17) What is a cyclone and discuss its impact on human lives?

18) Describe the structural and nonstructural approaches to disaster mitigation.

19) Define tsunamis and their various causes and consequences.

20) What are the major flood-prone regions of India?

2.11. Multiple-choice questions (MCQs)

1) What is the primary cause of tsunamis?

a) Volcanic eruptions

b) **Earthquakes**

c) Hurricane winds

d) Tidal forces

2) Which of the following is a natural disaster associated with extremely cold temperatures?

a) Tsunami

b) **Blizzard**

c) Heatwave

d) Tornado

3) Which natural disaster is characterized by the sudden movement of rock and soil down a slope?

a) Earthquake

b) Tsunami

c) **Avalanche**

d) Tornado

4) Which weather phenomenon involves rotating columns of air extending from a thunderstorm to the ground?

a) Hurricane

b) **Tornado**

c) Typhoon

d) Cyclone

5) What is the primary cause of a drought?

a) Excessive rainfall

b) **High temperatures**

c) Low atmospheric pressure

d) Strong winds

6) Which natural disaster is caused by the shifting of tectonic plates beneath the Earth's surface?

a) Hurricane

b) **Earthquake**

c) Tornado

d) Flood

7) Which natural event is responsible for the creation of hot springs and geysers?

a) Earthquake

b) **Volcanic activity**

c) Tsunami

d) Tornado

8) What is the primary cause of wildfires in forests?

a) Flooding

b) **Lightning**

c) Hailstorms

d) Tsunamis

9) Which geological process results in the formation of caves and sinkholes?

a) **Erosion**

b) Volcanic eruption

c) Earthquakes

d) Landslide

10) Which natural disaster is associated with the sudden collapse of a snowpack on a mountain slope?

a) **Avalanche**

b) Tornado

c) Blizzard

d) Earthquake

11) Which natural disaster is associated with the melting and movement of glacial ice?

a) Tsunami

b) Avalanche

c) **Glacier burst**

d) Cyclone

12) What is the primary cause of a volcanic eruption?

a) Earthquake

b) Tsunami

c) **Tectonic activity**

d) Hailstorm

13) Which atmospheric condition is responsible for the formation of sandstorms in desert regions?

a) Low humidity

b) High pressure

c) **Strong winds**

d) Tornados

14) Which natural event is caused by the gravitational pull of the Moon and the Sun on Earth's oceans?

a) Tsunami

b) Tornado

c) **Tidal wave**

d) Hurricane

15) What is the primary cause of a heatwave?

a) Low atmospheric pressure

b) High humidity

c) **Intense solar radiation**

d) Strong winds

16) Which natural disaster involves the rapid freezing of water on surfaces, creating a layer of ice?

a) Tsunami

b) Avalanche

c) **Blizzard**

d) Tornado

17) Which geological process is responsible for the formation of mountain ranges?

a) Volcanic activity

b) Erosion

c) **Plate tectonics**

d) Tsunami

18) Which natural disaster is associated with the collapse of underground caverns?

a) Earthquake

b) **Sinkhole**

c) Landslide

d) Avalanche

19) Which atmospheric condition is associated with the formation of cyclones in tropical regions?

a) **Low pressure**

b) High humidity

c) Strong winds

d) Tornado

20) What is the primary cause of a mudslide or mudflow in hilly terrain?

a) Volcanic activity

b) **Heavy rainfall**

c) Earthquake

d) Tornado

21) Which natural disaster is caused by the overflow of rivers or other water bodies onto land?

a) Tsunami

b) **Flood**

c) Hurricane

d) Blizzard

22) Which organization is responsible for issuing tsunami warnings in the Pacific Ocean region?

a) **National Oceanic and Atmospheric Administration (NOAA)**

b) United Nations Development Programme (UNDP)

c) International Red Cross

d) World Health Organization (WHO)

23) The term "drought" is an example of which type of disaster?

a) Geophysical

b) **Climatological**

c) Meteorological

d) Hydrological

24) The state of J&K is classified in which seismic zone?

a) Zones II and III

b) Zones V and II

c) Zones III and V

d) **Zones IV and V**

25) The October 8, 2005 earthquake that resulted in death of more than 80,000 people was of which magnitude?

a) 5.6

b) 6.6

c) **7.6**

d) 8.6

26) The scientific study of earthquakes is called:

a) Seismograph

b) **Seismology**

c) Both a and b

d) None of the above

27) What is the name of the instrument used for measuring earthquake intensity?

a) **Modified Mercalli Scale**

b) Metrological scale

c) Vernier scale

d) Barometer

28) The primary result of earthquakes is:

a) Building and bridge collapse

b) Rupture of water and gas pipelines

c) Change in the course of the river and creation of new islands

d) **All of the above**

29) What are the major consequences of tropical cyclones?

a) Fierce winds

b) Heavy rain

c) Storm surges

d) **All of the above**

30) The major man-made causes of floods are:

a) Deforestation

b) Siltation

c) Bursting of a dam

d) **All of the above**

31) What percentage of Indian land is prone to earthquakes?

a) 52%

b) **58%**

c) 59%

d) 60%

32) A point on the surface of the Earth vertically above the focus is known as:

a) **Epicenter**

b) Hypocenter

c) Special focus

d) None of the above

33) Which instrument is used for recording the occurrences of an earthquake?

a) Richter Scale

b) Seismology

c) **Seismograph**

d) None of the above

34) Which area in the world witnesses the highest seismic earthquake activity?

a) Mid-continental belt

b) **Circum Pacific belt**

c) Mid-Atlantic belt

d) All of the above

35) Which type of disaster is most prominent in India?

a) **Flooding**

b) Droughts

c) Cyclones

d) Earthquakes

36) Which volcanic eruption is responsible for raising the temperature of the Northern Hemisphere by 1.2°C?

a) Bhuj

b) Etna

c) **Krakatoa**

d) Mt. Pinatubo

37) Which are the major controlling systems for the reduction of volcanic disasters?

a) Prediction for early warning of eruption

b) Timely evolution and relief work

c) **Both a and b**

d) None of the above

38) Which techniques are used for the prediction of early warning signs of eruption?

a) Electric distance measures

b) Tilt meters

c) GPS and satellite inferring radar

d) **All of the above**

39) Which of the following disasters can be triggered by an earthquake?

a) Tsunami

b) Intense ground shaking

c) A landslide

d) **All of the above**

40) Which of the following statements is false?

a) Earthquakes occur in plate boundaries

b) **The time and location of most major earthquakes can be predicted several days in advance**

c) Earthquakes can be caused by normal, reverse and strike-slip faulting

d) P-waves travel faster than both S-waves and Surface waves

41) The center of the cyclone is characterized by

a) High pressure

b) **Low pressure**

c) Very high pressure

d) All of the above

42) Warnings of cyclonic disasters are reported in:

a) Newspapers

b) Radio networks

c) The television

d) **All of the above**

43) The major effects of flooding are:

a) Physical damage

b) Loss of biodiversity

c) Scarcity of water supply

d) **All of the above**

44) Tropical cyclones, droughts, and wildfires are:

a) Industrial hazards

b) Man-induced hazards

c) **Meteorological hazards**

d) Hydrological hazards

45) Earthquakes are:

a) **Geophysical disasters**

b) Hydrological disasters

c) Meteorological disasters

d) Compound disasters

46) Mount Kilimanjaro is the best example of an:

a) Active volcano

b) **Dormant volcano**

c) Extinct volcano

d) None of the above

47) Which of the following leads to earthquakes?

a) Movement of plates

b) Nuclear explosions

c) Extraction of minerals

d) **All of the above**

48) Tsunamis can only occur during:

a) Evenings

b) Afternoons

c) **Anytime, of day or night**

d) Mornings

49) The term Tsunami is coined from a:

a) Chinese term

b) Indian term

c) German term

d) **Japanese term**

50) Landslides occur because of:

a) The intensity of rainfall

b) Steep slopes

c) Deforestation leading to soil erosion

d) **All the above**

51) International Tsunami Information Centre is in:

a) **Honolulu**

b) Goa

c) Jakarta

d) Pondicherry

52) What is the cause of tidal waves?

a) Wind

b) **Gravitational force of the Moon on the sea**

c) Gravitational force of the Sun on the sea

d) Gravitational force of the Earth on the sea

53) The center of a cyclone is a calm area and is called the__________ of the storm.

a) **Eye**

b) Focus

c) Center

d) Radius

54) A cyclone is called a________in the American continent.

a) **Hurricane**

b) Typhoon

c) Tornado

d) Thunderstorm

55) A cyclone is called a____________in the Philippines and Japan.

a) Hurricane

b) **Typhoon**

c) Tornado

d) Thunderstorm

56) Which one of the following is NOT recommended during lightning or thunderstorms?

a) Taking shelter in a bus or car

b) In a forest, seeking shelter in a low area under a thick growth of small trees

c) **Taking shelter under an isolated tree**

d) Getting away from tractors and other metal farm equipment

57) The eye of a cyclone is

a) Cloudy and highly stormy

b) Full of strong winds

c) **Calm and cloudless**

d) A site with high atmospheric pressure

58) A flash flood is a flood that:

a) Is caused by heavy rain rather than by the flooding of a river

b) Occurs in urban areas

c) **Occurs suddenly and unexpectedly and for a short duration**

d) Is caused by the blocking of drains

59) Which of the following is an environmental consequence of floods?

a) Dispersal of weed species

b) Erosion of soil

c) Release of pollutants into waterways

d) **All of the above**

60) On which date did a tsunami occurr in Indonesia?

a) 24 April

b) 26 March

c) **26 December**

d) 27 December

61) The term cyclone means:

a) Coils of a tiger

b) **Coils of a snake**

c) Coils of a lion

d) Coils of a pig

62) The Earth's crust is made up of huge slabs, which are called:

a) Crust

b) **Plates**

c) Rock

d) Magma

63) When magma reaches the surface of the Earth, it is called:

a) **Lava**

b) Crater

c) Fissure

d) Clay

64) The magnitude of energy released by an earthquake is usually measured on the Richter scale which ranges between:

a) 0 and 7

b) 0 and 8

c) **0 and 9**

d) 0 and 12

65) Which volcanic belt is commonly called the "Ring of Fire"?

a) The Mid-Continental Belt

b) **The Circum-Pacific Belt**

c) The Mid-Atlantic Belt

d) The Circum-Continental Belt

66) The Latur earthquake occurred on:

a) **September 30, 1993**

b) November 30, 1993

c) April 21, 1993

d) April 15, 1993

67) The Bhuj earthquake occurred on:

a) January 23, 2001

b) **January 26, 2001**

c) January 31, 2001

d) January 29, 2001

68) Tsunamis are also known as:

a) Tectonic sea waves

b) **Seismic sea waves**

c) Platonic sea waves

d) Magnetic waves

69) The Indian Tsunami Early Warning Centre (ITEWC) established at the Indian National Centre for Ocean Information Sciences is located in:

a) Chennai

b) **Goa**

c) Kochi

d) Hyderabad

70) The Richter scale is used to measure?

a) Density of liquids

b) Speed of airplanes

c) **Intensity of earthquakes**

d) Depth of mines

71) Landslides occur because of:

a) Rainfall

b) Steep slopes

c) Deforestation

d) **All of the above**

72) What is a cyclone?

a) A low-pressure system with clockwise winds in the northern hemisphere

b) A high-pressure system with anticlockwise winds in the northern hemisphere

c) **A low-pressure system with anticlockwise winds in the northern hemisphere**

d) A high-pressure system with clockwise winds in the northern hemisphere

73) From the following, which one is a natural hazard?

a) Earthquake

b) Cyclone

c) Landslide

d) **All of the above**

74) Mt. Etna is a:

a) Extinct volcano

b) Dormant volcano

c) **Active volcano**

d) None of the above

75) Which of the following is a disaster mitigation strategy?

a) **Constructing cyclone shelters**

b) Getting loans from banks

c) Providing cheap electricity

d) Providing school uniforms to children

3

Human-induced Disasters

Human-induced disasters are the events or crises caused directly or indirectly by human activities. These disasters can have significant environmental, social and economic impacts. While some human-induced disasters may result from unintentional actions or negligence, others may arise from deliberate human activities or decisions. These hazards can have diverse origins, including industrial processes, urbanization, agriculture and resource extraction. Unlike natural disasters, which are triggered by natural forces, man-made disasters are preventable and often stem from human negligence, error or malicious intent. These disasters can have widespread and often severe consequences for ecosystems, communities and economies. Some common types of human-induced disasters that result from human activities and have a huge impact on the environment include biological disasters or biohazards, fires (urban or domestic fires), nuclear (weapon), chemical (oil spills) and pollution (air water, soil and noise pollution), transportation (road, rail and water accidents), technological hazards, campus shooting and bomb threats. Communal riots, terrorism, militancy, ethnic conflicts, plight of refugees, internally displaced persons, epidemics, industrial accidents, etc., cause a significant loss of life and property. Natural disasters cannot be completely predicted and controlled, while human-induced disasters may be controlled through proper planning and preparedness. Human-induced disasters pose significant threats to human safety, environmental sustainability and socioeconomic stability. Humans contribute to climate change, leading to hazards such as extreme weather events, sea-level rise and shifts in disease patterns. Pollution from industrial, agricultural and urban sources contributes to air and water pollution, posing risks to human health and ecosystems. Addressing the root causes of these disasters requires concerted efforts to improve risk management practices, enhance regulatory oversight and promote community resilience. Understanding these hazards involves assessing their potential impacts on human health, ecosystems and infrastructure, as well as identifying and implementing measures to prevent and respond to these hazards and developing effective risk management strategies, mitigating their impacts. This may involve regulatory frameworks, land-use planning, disaster preparedness, technological innovations and public awareness campaigns to mitigate the impacts of human-induced hazards. By implementing proactive mitigation strategies, including risk assessment, preparedness planning, regulatory measures and public engagement, societies can reduce the likelihood and severity of human-induced disasters and build more resilient and sustainable communities for the future.

3.1. Transportation disasters

3.1.1. *Meaning and concept*

Transportation disasters encompass a wide range of incidents that occur in various forms of transportation. These disasters are associated with mechanical means of transportation, automobiles on roads, rails and trains; boats, fishing boats, ferries, submarines, passenger ships, vessels transporting migrants and tankers; and aircraft, helicopters, airships and balloons are included in this category. Disasters related to rails and roads can be caused by a variety of kinetic, thermal and/or chemical risks, including collisions, mechanical failure, weather, fires, explosions and/or human error. One well-known transportation tragedy was the sinking of the Titanic in 1912. The onset of transportation disasters may be sudden (e.g. an airplane crash) or it may take few hours (such as when a ship sinks). Secondary explosions, poisonous exposures, environmental contamination, fires and building collapse are all possible outcomes of transportation-related disasters. Big events, like the World Trade Center explosion in 2001, are known to have long-term effects on the economy of a region. Critical infrastructures, including electricity, commerce, healthcare facilities and communications can also be affected by transportation disasters.

3.1.2. *Causes of transportation disasters*

Transportation disasters can be caused by a variety of factors. The following are some common causes of transportation disasters:

1) *Human error*: one of the leading causes of transportation disasters is human error, including mistakes made by pilots, train operators, ship captains and drivers. Fatigue, distraction and poor decision-making can all contribute to accidents. In addition, factors such as substance abuse, medical conditions or inadequate training can also contribute to transportation disasters.

2) *Mechanical and infrastructure failures*: malfunctions or failures of critical components in vehicles, aircraft or ships can lead to catastrophic accidents. These failures can result from design flaws, poor maintenance, manufacturing defects or unexpected events such as natural disasters; these can all play a role in mechanical failures. In addition, collapses of bridges, tunnels, or other transportation infrastructure can lead to major disasters.

3) *Weather conditions*: adverse weather, such as fog, heavy rain, snow, or ice, can significantly impair visibility and create hazardous driving, flying or sailing conditions, increasing the risk of accidents.

4) *Terrorism*: deliberate acts of terrorism, such as bombings or hijackings, can result in transportation disasters with devastating consequences for passengers and the public at large.

5) *Inadequate safety regulations and oversight*: weak or poorly enforced safety regulations, as well as insufficient oversight of transportation systems, can contribute to an increased risk of accidents and disasters.

3.1.3. *Effects of transport disasters*

Transport disasters can have far-reaching impacts on individuals and communities. The following are some of the key effects of transport disasters:

1) *Loss of life and injury*: one of the most immediate and devastating effects of transport disasters is the loss of life and the occurrence of injuries, whereby victims and their families may experience severe and long-lasting effects from disasters.

2) *Infrastructural damage*: these disasters result in significant damage to infrastructure, including roads, bridges, railways, airports and seaports, disrupting transportation networks and impeding the movement of goods and people. In addition, the disruption of essential services, such as emergency response, healthcare and utilities, further complicates the response and recovery efforts.

3) *Economic impact*: transport disasters can have substantial economic repercussions, including direct costs associated with rescue and recovery efforts, as well as indirect costs related to business interruptions, loss of productivity and decreased tourism.

4) *Environmental consequences*: accidents involving hazardous materials or fuel spills can lead to environmental contamination, posing risks to ecosystems, wildlife and public health.

5) *Psychological and emotional trauma*: survivors of transport disasters, as well as witnesses and first responders, can experience psychological trauma, including post-traumatic stress disorder (PTSD) and other mental health challenges.

3.1.4. Control measures of transportation disasters

To mitigate the impact of transportation disasters, various control measures can be implemented across different modes of transportation. These measures aim to enhance safety, reduce the risk of accidents and improve emergency response capabilities. Some key control measures include:

– *Stringent safety regulations*: implementing and enforcing strict safety regulations for all modes of transportation, including road, rail, maritime, and aviation, to ensure compliance with safety standards and protocols.

– *Advanced training and education*: providing comprehensive training programs and launching public awareness initiatives for pilots, drivers and railway personnel to educate travelers to enhance their skills about safety procedures, the importance of adhering to regulations and the need for vigilance while using transportation services and ability to respond effectively to emergencies.

– *Technological innovations*: these innovations embrace technological advancements such as collision avoidance systems, automatic braking systems, real-time monitoring and communication technologies to improve safety and reduce the likelihood of accidents.

– *Regular maintenance and inspections*: instituting regular maintenance schedules and thorough inspections of vehicles, aircraft, ships and infrastructure to identify and address potential safety hazards and mechanical issues. Investing in the maintenance, repair and modernization of transportation infrastructure to ensure its reliability, resilience and ability to withstand various environmental and operational challenges.

– *Emergency response preparedness*: developing and regularly rehearsing emergency response plans to ensure swift and effective reactions to transportation disasters, including coordination among first responders, medical facilities and relevant authorities.

– *Weather monitoring and alerts*: utilizing advanced weather monitoring systems to track and predict adverse weather conditions, enabling proactive measures to be taken to minimize the impact on transportation.

3.1.5. *Road safety tips*

3.1.5.1. *Dos*

– *Follow traffic laws*: obey speed restrictions, traffic signals and signage. Following the law lowers the chance of accidents and encourages safe driving.

– *Wear seatbelts*: always wear seatbelts while driving or riding in a vehicle, regardless of your position. Seatbelts significantly reduce the risk of injury and save lives in the event of a collision.

– *Maintain safe distance*: keep a safe distance between your vehicle and the vehicle ahead of you to allow for sufficient braking distance and reaction time. Maintain a three-second gap under normal driving conditions and increase the gap in adverse weather or road conditions.

– *Use turn signals*: signal your intentions by using turn signals or indicators when changing lanes, making turns or merging into traffic. Clear communication with other drivers helps prevent misunderstandings and collisions.

– *Stay alert and focused*: avoid distractions while driving, such as texting, talking on the phone, eating or adjusting electronic devices. Stay focused on the road and maintain awareness of your surroundings to anticipate potential hazards.

– *Check blind spots*: before changing lanes or making turns, check your blind spots by using rear-view and side mirrors and conducting shoulder checks. Blind spot monitoring systems can also assist in detecting vehicles in adjacent lanes.

– *Observe right of way*: yield the right of way to pedestrians, cyclists and other vehicles as required by traffic laws. Respect other road users and exercise caution at intersections, crosswalks and merging lanes.

– *Keep vehicle well maintained*: regularly inspect and maintain your vehicle to ensure the proper functioning of brakes, tyres, lights and other essential components. Address any mechanical issues promptly to prevent accidents due to vehicle malfunctions.

– *Drive defensively*: anticipate potential hazards and react proactively to prevent accidents. Be prepared to adjust your driving behavior in response to changing road conditions, weather and traffic patterns.

– *Follow emergency procedures*: in the event of an accident or emergency, follow established procedures such as moving vehicles to a safe location, exchanging contact and insurance information, and seeking medical assistance for injuries.

3.1.5.2. *Don'ts*

– *Don't drive under the influence*: never drive under the influence of alcohol, drugs or medications that impair your ability to operate a vehicle safely. Impaired driving significantly increases the risk of accidents and endangers lives.

– *Don't speed*: avoid speeding or driving recklessly, as excessive speed reduces reaction time and increases the severity of accidents. Obey designated speed limits and modify your speed by the state of the road.

– *Don't engage in road rage*: avoid aggressive driving behaviors such as tailgating, honking excessively, making rude gestures, or engaging in confrontations with other drivers. Keep calm and maintain composure to prevent road rage incidents.

– *Don't drive while fatigued*: avoid driving when you are fatigued or drowsy, as it impairs your reaction time and decision-making ability. Take breaks during long trips and get an adequate amount of rest before driving.

– *Don't use electronic devices*: refrain from using electronic devices such as mobile phones, tablets or GPS systems while driving, as they distract your attention from the road and increase the risk of accidents.

– *Don't run red lights or stop signs*: always obey traffic signals and stop signs. Running red lights or stop signs increases the likelihood of collisions with crossing traffic or pedestrians.

– *Don't drive aggressively*: avoid aggressive driving behaviors such as weaving in and out of traffic, cutting off other vehicles, or engaging in road rage. Drive defensively and courteously to promote safety for yourself and others.

– *Don't drive distracted*: refrain from engaging in activities that distract you from driving, such as eating, grooming or using electronic devices. Maintain focus on the road and avoid distractions to ensure safe driving.

3.2. Urban fires

3.2.1. *Meaning and concept*

– Urban fires occur primarily in cities and towns with the potential to rapidly spread to adjoining structures. Urban fires refer to large-scale fires that occur in densely populated areas, often affecting buildings, vehicles, school infrastructure and communities in urban areas. These fires can be caused by various factors, including accidents, arson, human negligence, carelessness, industrial incidents or natural disasters, leading to loss of life, property damage, displacement of populations and significant challenges for emergency responders. Accidental causes, such as careless handling of flammable materials, misuse of appliances or negligence in following fire safety guidelines, can contribute to urban fires. In addition, storage of combustible materials, such as paper, cardboard, or flammable liquids, near heat sources or ignition points can increase the risk of fires. These fires can have significant consequences due to the concentration of people, property and resources in urban environments. Urban fires may occur in residential, commercial, industrial or mixed-use areas and can result from various causes, including accidents, electrical faults, arson or natural disasters such as earthquakes or wildfires spreading into urban areas. These hazards pose unique challenges compared to fires in rural or remote areas due to the high population density, complex built environment, limited access for emergency vehicles and potential for rapid fire spread between buildings. Displaced residents may face challenges in rebuilding their lives, and communities may experience changes in demographics and social dynamics. The intensity and scale of urban fires can vary widely, ranging from small structural fires contained within a single building to large conflagrations engulfing multiple structures and requiring extensive firefighting efforts. Efforts to prevent and mitigate urban fires involve various strategies, including fire prevention education, building code enforcement, fire safety inspections, installation of fire detection and suppression systems, emergency preparedness planning and coordinated response by fire departments and emergency services. Effective urban fire management also involves collaboration with urban planners, architects, engineers and community stakeholders to design and construct fire-resistant buildings, improve access for emergency responders and enhance overall urban resilience to fire hazards.

3.2.2. *Causes of urban fires*

Urban fires can be caused by a variety of factors, and their origins may range from accidental occurrences to deliberate acts. Here are some common causes:

– *Electrical malfunctions*: electrical issues, such as faulty wiring, overloaded circuits or electrical equipment malfunctions, can lead to short circuits and sparks, resulting in fires. In urban areas, where buildings are closely spaced and interconnected, electrical fires can spread rapidly.

– *Cooking and heating accidents*: unattended cooking, overheating of cooking equipment or grease fires in kitchens are common causes of urban fires, particularly in residential buildings. Malfunctioning or improperly maintained heating systems, including furnaces, boilers and space heaters, can cause fires in both residential and commercial buildings, especially during the colder months.

– *Smoking*: discarded cigarette butts, matches or the improper disposal of smoking materials can ignite fires, especially in areas where flammable materials are present.

– *Arson*: deliberate acts of arson, where someone intentionally sets fire to buildings or property, can lead to destructive urban fires. Arson may be motivated by various factors, including vandalism, insurance fraud or criminal intent.

– *Industrial accidents*: in areas with industrial facilities, accidents such as chemical spills, explosions or equipment malfunctions can result in fires that pose a threat to surrounding urban areas.

– *Gas leaks*: leaking natural gas or propane can create explosive and flammable atmospheres, leading to fires and explosions. Gas leaks may occur in residential, commercial or industrial settings.

– *Human error*: accidental causes, such as careless handling of flammable materials, misuse of appliances or negligence in following fire safety guidelines, can contribute to urban fires. In addition, storage of combustible materials, such as paper, cardboard, or flammable liquids, near heat sources or ignition points can increase the risk of fires.

– *Fireworks*: improper use of fireworks, particularly in densely populated areas, can lead to accidental fires. Sparks or embers from fireworks can ignite flammable materials on buildings or in open spaces.

– *Natural disasters*: natural disasters, such as earthquakes, hurricanes or wildfires, can indirectly lead to urban fires. For example, earthquakes may damage gas lines, sparking fires, while wildfires can spread to urban areas.

3.2.3. *Effects of urban fires*

Urban fires can have far-reaching and devastating impacts on communities, economies and the environment, depending on factors such as the scale of the fire, the density of the urban area and the effectiveness of emergency response efforts; some common effects of urban fires are as follows:

– *Loss of life*: urban fires can lead to the tragic loss of human lives. The immediate threat to residents, firefighters and other emergency responders is significant during the fire, and the long-term health impacts from smoke inhalation can also contribute to casualties.

– *Infrastructure property damage*: buildings, homes and infrastructure are often severely damaged or destroyed during urban fires. The economic cost of rebuilding or replacing structures can be substantial, impacting individuals, businesses and local governments. People living in the affected area may be forced to evacuate due to the fire, leading to temporary or permanent displacement. This can strain emergency shelters, disrupt communities and create challenges for local authorities in managing the aftermath. Additionally, urban fires can damage critical infrastructure such as power lines, water supply systems and transportation networks.

– *Economic impact*: the economic consequences of urban fires include the direct costs of property damage, business interruptions and the loss of livelihoods. The recovery process can take years, in addition, the costs associated with property damage and losses are often covered by insurance. However, the widespread impact of urban fires can lead to increased insurance premiums for individuals and businesses in the affected region.

– *Environmental and health consequences*: the environmental and health consequences of urban fires can persist long after the flames are extinguished. Urban fires release pollutants into the air, soil and water, contributing to environmental degradation. The combustion of building materials and hazardous substances can lead to air pollution, soil contamination and damage to ecosystems. Smoke inhalation and exposure to toxic substances released during urban fires can cause respiratory problems, exacerbate existing health conditions and have long-term health effects on both residents and emergency responders.

– *Social and psychological effects*: the trauma and stress associated with urban fires can have lasting psychological impacts on affected individuals and communities. The loss of homes, possessions and a sense of security can lead to mental health challenges. The disruption caused by urban fires can strain social networks and community cohesion.

3.2.4. *Mitigation measures of urban fires*

Mitigating the impacts of urban fires involves a combination of preventive measures; some of the effective mitigation measures are discussed as follows:

– *Urban planning and zoning*: it implements and enforces building codes and zoning regulations that promote fire-resistant construction, adequate spacing between structures and proper placement of buildings in urban areas.

In addition, it engages communities in developing fire-adapted plans that consider local conditions and risks and encourages collaboration among residents, local government and emergency services to collectively address fire hazards.

– *Fire-resistant construction materials*: it encourages the use of fire-resistant materials in building construction, such as non-combustible roofing, siding and windows. In addition, it upgrades and maintains critical infrastructure, including water supply systems, roads and communication networks, to support firefighting efforts and emergency response. This will help in reducing the vulnerability of fires.

– *Public education and awareness*: it conducts public education campaigns to raise awareness about fire safety, prevention and evacuation procedures. It educates residents about the risks associated with various fire hazards and the importance of responsible behavior. It also regularly practices drills to ensure preparedness.

– *Early warning systems*: implement and maintain early warning systems, such as fire detection sensors, alarms and community alert systems, to provide timely notifications to residents and emergency responders.

– *Emergency evacuation plans*: it develops and communicates clear emergency evacuation plans that outline evacuation routes, assembly points and procedures for residents to follow during a fire.

– *Firebreaks maintenance*: it establishes and maintains physical firebreaks and green spaces such as cleared areas or roads to create barriers that can help control the spread of fires. To ensure their efficacy, regular maintenance is necessary.

– *Emergency services training*: it includes training emergency service personnel in urban firefighting techniques, including strategies for high-rise buildings and densely populated areas. It ensures that firefighting equipment is up-to-date and well maintained.

– *Insurance and risk assessment*: it encourages property owners and businesses to carry adequate insurance coverage. It also conducts risk assessments to identify and address vulnerabilities in buildings and infrastructure.

3.3. Biological disasters

3.3.1. *Meaning and nature*

Biological disaster or a hazard is the transmission of diseases among living organisms like humans, animals and plants caused by microorganisms such as bacteria, toxins or viruses. It is defined as the catastrophic consequences resulting from the widespread spread of a particular type of living creature, which has the potential to spread viruses, diseases or infestations of plant, animal or insect life on an epidemic or pandemic scale. These are the catastrophic events caused by the outbreak of infectious diseases or the release of harmful biological agents. These events have the potential to disrupt ecosystems, endanger lives and have long-lasting impacts on public health and the environment. Biological disasters or biohazards refer to the biological substances or organic matters that originate from living organisms such as parasites, bacteria, fungi and viruses, and may also evolve due to the release of harmful biological agents from laboratories or research facilities. Medical waste that threatens the health of living organisms (especially humans) may cause loss of life, illness or other health effects, loss of services and livelihoods, disturbance of social and economic stability, or destruction of the environment. These disasters can be natural, accidental or intentional. The transmission of these hazards can occur through various pathways, including direct contact, ingestion, inhalation or vector-borne transmission, for example, cholera, H1N1virus and Covid-19.

Biohazards involve the release of biological agents or the occurrence of infectious diseases on a large scale, causing significant harm to human health, ecosystems and animals. Strict safety measures and protocols are essential to prevent such incidents. Factors such as climate change, urbanization and antimicrobial resistance can impact the behavior of these hazards. These organisms have the potential to reproduce and evolve. Certain living organisms produce toxins that can be harmful when ingested or come into contact with humans, animals and insects. Toxins can be found in various sources, such as poisonous plants, venomous animals and certain species of bacteria and fungi. Some biological hazards, such as pollen, certain foods and animal dander, can trigger allergic reactions in susceptible individuals. Allergens can lead to respiratory issues, skin rashes and other allergies. It includes bacteria, viruses, fungi and parasites that have the potential to cause infectious diseases. Examples of infectious agents include

Salmonella bacteria, Influenza virus, Aspergillus fungi, and Plasmodium parasites. In India, the Ministry of Health and Family Welfare (MoHFW) is the Nodal ministry for making decisions, managing outbreaks, serving as an advising council and providing emergency medical relief. According to the constitution, health is a matter for the state. Therefore, state governments bear the primary duty of responding to biological disasters. The primary organization for investigating outbreaks in the nation is the National Institute of Communicable Diseases (NICD). Furthermore, the NICD/Indian Council of Medical Research (ICMR) provides laboratory support, research and teaching/training.

– *Epidemic*: a localized or regional outbreak of infectious disease that affects a large group of individuals inside a population or in a community, simultaneously. It has an infectious origin and occurs in the aftermath of a particular disaster, for example, cholera, plague and Zika virus in various regions.

– *Pandemic*: it is a global outbreak of infectious disease that spreads quickly throughout a wide region, such as a continent, or perhaps worldwide, of illnesses and pestilences that are now prevalent, newly developing or re-emerging diseases such as Covid-19 and H1N1.

3.3.1.1. *Biological warfare (BW)*

"Germ warfare" is another term for the use of biological weapons or infectious agents, such as fungi, bacteria, and viruses to intent or harm humans, animals, insects and plants as an act of war. The frequent employment of biological poisons, toxins and/or infectious agents, such as bacteria, viruses and fungus, to kill or harm people, animals and insects or plants as a weapon of war is known as BW or germ warfare. Biological weapons, also referred to as "bioweapons", "biological threat agents", or "bio-agents", are living organisms (such as viruses, which are not generally considered to be living objects) that reproduce within their host victims. Another class of biological weapons is deemed to be entomological (insect) warfare. Contrary to chemical and nuclear warfare, which together with biological warfare constitute the military acronym for nuclear, biological and chemical warfare employing weapons of mass destruction (NBC), this form of warfare is distinct. The MoHFW collaborates with the Ministry of Home Affairs (MHA) to oversee biological warfare as the nodal ministry. MHA is responsible for assessing threat assessments, putting in place deterrents and contributing intelligence. The Epidemic Act of 1987 gives the Indian government's state and central branches the means to enact laws and conduct other actions that will help them control biological disasters.

3.3.2. *NDMA guidelines for management of biohazards*

1) It is necessary to perform vulnerability analysis and risk assessment at both the macro- and micro-levels for current diseases that have the potential to become epidemics, as well as for newly developing and re-emerging diseases that have the potential to spread to others.

2) When exposure is suspected, the affected persons shall be quarantined and put under monitoring for any atypical or typical signs and symptoms taking shape during the period of observation.

3) Health professionals associated with such investigations will have adequate protection and adopt recognized universal precautions.

4) Healthcare workers shall be equipped with gloves, impermeable gowns, N-95 masks or powered air-purifying respirators; they must sanitize their hands before donning PPE for patient contact.

5) With due consideration to the social, ethnic and religious issues involved, utmost care will be taken in the disposal of dead bodies.

6) Necessary awareness regarding the importance of personal hygiene and measures should be given to the community; this includes the provision of washing, cleaning and bathing facilities, avoiding overcrowding in sleeping quarters, etc.

7) Attention is to be given to developing infrastructural facilities in terms of trained human resources, mobility, connectivity, knowledge enhancement and scientific up-gradation for all the stakeholders concerned with the management of biological disasters.

8) Biological disaster-related education and awareness shall be given in several informal languages. To increase awareness, district-level employees will also develop simple exercise models.

9) For early responders or the general public, toll-free emergency numbers and an award system for supplying critical information about any impending biological calamity will be useful.

10) Those families subjected to bereavement of their near and dear ones would also reflect higher psycho-social morbidity. MoHFW, through its

mental health institutions and NGOs, would provide sufficient psycho-social care.

3.3.3. *Impact of biological hazards*

The impact of biological hazards can be far-reaching, affecting various aspects of human society, the economy and the environment:

1) *Public health impacts*: biological hazards are major contributors to the global burden of disease. Infectious diseases caused by biological hazards can lead to illness, disability and mortality, posing significant challenges to healthcare systems worldwide. The trauma and stress resulting from biological disasters can have long-lasting effects on the mental well-being of the affected people.

2) *Economic consequences*: outbreaks of infectious diseases can have substantial economic repercussions. These include healthcare costs, loss of productivity due to illness, overall disruptions to trade and tourism and expenses associated with disease prevention and control measures.

3) *Environmental effects*: biological hazards can also impact ecosystems and biodiversity. Invasive species, pathogens that affect wildlife and diseases that impact crops can have profound ecological consequences.

4) *Social disruption*: outbreaks of infectious diseases can lead to social disruption, fear, stigma and discrimination. Communities and individuals may face challenges in maintaining social and economic activities during disease outbreaks.

3.3.4. *Response and management*

Management of biological hazards involves various approaches that encompass prevention, mitigation, surveillance, preparedness and response. The following are the important mitigation measures for biological hazards or disasters:

1) *Preventive measures*: preventing the emergence and spread of biological hazards involves measures such as vaccination, biosecurity practices in agriculture, food safety protocols and environmental management to reduce exposure to disease vectors.

2) *Surveillance and early detection*: integrated disease surveillance systems for monitoring infectious diseases and other biological hazards play a crucial role in early detection, allowing for timely intervention to prevent further spread.

3) *Risk communication and awareness*: informing the public, healthcare professionals and policymakers about biological hazards and appropriate preventive measures is essential for promoting awareness and facilitating a coordinated response to potential threats. Continual learning and awareness are essential for effective disaster preparedness. Working together with global partners enhances response capabilities and resilience.

4) *Outbreak response*: prompt and coordinated responses to biological hazard outbreaks are vital to containing the spread of diseases. This involves deploying healthcare resources, implementing quarantine measures and conducting contact tracing to limit transmission.

5) *Proper maintenance*: proper maintenance of sewage pipelines and safe water supply can go a long way in preventing biological disasters and epidemics of waterborne origin such as cholera, hepatitis, diarrhoea and dysentery.

3.4. Nuclear (radiation) disasters

3.4.1. *Meaning and concept*

Nuclear disasters are catastrophic events that involve the release of dangerous radioactive materials from nuclear facilities into the surroundings, leading to widespread environmental contamination, health risks and social and economic disruptions. These disasters can result from accidents, meltdowns or intentional acts. Instances of human error have played a significant role in various nuclear disasters, leading to catastrophic outcomes. Natural disasters like earthquakes, volcanoes and tsunamis have also triggered nuclear disasters. In addition, failures in critical systems and infrastructure have also contributed to the occurrence of nuclear disasters throughout history. These disasters are often the result of complex interactions between technological failures, human error, natural disasters and inadequate safety measures. Addressing the root causes of nuclear disasters requires comprehensive efforts to strengthen nuclear safety standards, enhance regulatory oversight, improve emergency preparedness and response capabilities and promote transparency and accountability in the nuclear industry. Ensuring an effective safety culture, rigorous risk assessment and continuous learning from past accidents are essential for preventing future nuclear disasters and safeguarding the well-being of communities and the environment.

Figure 3.1. *Nuclear hazards. For a color version of this figure, see www.iste.co.uk/akbar/disastermanagement.zip*

3.4.2. *Causes of nuclear disasters*

Here are some common causes of nuclear disasters:

1) *Technological failures*: malfunctions or failures in nuclear reactors, cooling systems, containment structures or other critical components of nuclear facilities can lead to nuclear disasters. Equipment failures may result from design flaws, manufacturing defects, inadequate maintenance or ageing infrastructure. Without proper functioning of safety systems and redundancies, nuclear reactors can experience accidents such as meltdowns, core breaches or uncontrolled releases of radioactive materials.

2) *Human error*: human error, including operational mistakes, procedural violations and mismanagement, can contribute to nuclear disasters. Errors by

operators, technicians, engineers or management personnel in operating nuclear facilities, conducting maintenance activities or responding to abnormal conditions can escalate into accidents with severe consequences. Inadequate training, supervision and communication among personnel can exacerbate the risk of human error, leading to nuclear disasters.

3) *Natural disasters*: natural disasters such as earthquakes, tsunamis, hurricanes, floods and tornadoes can trigger nuclear disasters by compromising the safety and integrity of nuclear facilities. Earthquakes and tsunamis, in particular, pose significant risks to nuclear reactors located in seismically active or coastal regions. Natural disasters can damage critical infrastructure, disrupt power supplies, disable safety systems and lead to loss of cooling capabilities, increasing the likelihood of nuclear accidents and radioactive releases.

4) *Terrorism and sabotage*: deliberate acts of terrorism, sabotage or malicious intent targeting nuclear facilities can cause nuclear disasters. Terrorist attacks, cyber-attacks, insider threats or sabotage by hostile actors can compromise the security, safety and operations of nuclear facilities, leading to catastrophic consequences. Efforts to prevent unauthorized access enhance security measures and safeguard nuclear materials are essential for mitigating the risk of intentional nuclear disasters.

5) *Radiation leaks and contamination*: nuclear disasters can result from leaks, spills or releases of radioactive materials from nuclear facilities, storage sites or transportation accidents involving radioactive materials. Contamination of air, water, soil and food supplies with radioactive isotopes poses significant risks to human health, ecosystems and agricultural production. Accidental releases of radioactive materials can occur during routine operations, maintenance activities, accidents or incidents involving malfunctioning equipment or inadequate containment measures.

6) *Lack of emergency preparedness and response*: inadequate emergency preparedness and response measures can exacerbate the consequences of nuclear disasters. Insufficient planning, coordination and resources for emergency response, evacuation, medical care and decontamination can impede efforts to mitigate the impacts of nuclear accidents and protect public health and safety.

3.4.3. *Some notable major nuclear disasters*

1) *Three Mile Island Accident (1979)*: the Three Mile Island nuclear accident occurred in Pennsylvania. A partial reactor meltdown released a small amount of radioactive gases, causing concerns about potential health impacts. While there were no immediate injuries, the incident led to increased scrutiny of nuclear power safety.

2) *Chernobyl Nuclear Disaster (1986)*: the Chernobyl nuclear disaster occurred at the Chernobyl Nuclear Power Plant in Pripyat, Ukraine, then part of the Soviet Union. A reactor exploded during a safety test, resulting in the release of a massive amount of radioactive material into the surroundings. It is the worst nuclear accident in recorded history and resulted in immediate deaths of two plant workers and caused long-term health issues, including cancer, for many others. The release of radioactive material contaminated a large area, forcing the evacuation of nearby towns and communities.

3) *Deepwater Horizon Oil Spill (2010)*: this occurred near the Gulf of Mexico, off the coast of Louisiana. and caused due to an offshore drilling rig explosion and subsequent oil spill. One of the worst environmental catastrophes in recorded history is the deepwater horizon oil spill. The explosion resulted in the deaths of 11 workers and the release of millions of barrels of oil into the Gulf of Mexico. The spill had devastating effects on marine life, coastal ecosystems and local economies.

4) *Fukushima Daiichi Nuclear Disaster (2011)*: the Fukushima nuclear disaster was triggered by a powerful earthquake and tsunami that damaged the Fukushima Daiichi Nuclear Power Plant, leading to meltdowns in three of its reactors. The Fukushima disaster resulted in the release of radioactive material, causing evacuations and long-term health concerns.

3.4.4. *Impact of nuclear disasters*

Nuclear disasters have profound and far-reaching impacts on human health, the environment and socioeconomic stability.

– The release of radioactive materials during nuclear accidents can result in immediate and long-term health consequences, including radiation sickness, cancer, genetic mutations and reproductive disorders. Contamination of air, water, soil and food supplies with radioactive isotopes poses risks to ecosystems, biodiversity and agricultural production.

– The displacement of populations, evacuation of communities, and loss of livelihoods due to nuclear disasters contribute to social disruption, economic losses and psychological trauma among affected individuals and communities.

– Cleanup and decontamination efforts following nuclear disasters require significant financial resources, technological expertise and long-term monitoring to mitigate the environmental and health impacts and restore affected areas to safe and habitable conditions. The legacy of nuclear disasters can persist for generations, underscoring the importance of stringent safety measures, robust regulatory oversight and transparent communication in preventing future accidents and protecting public health and safety.

– Nuclear disasters have far-reaching consequences, causing severe damage and destruction to ecosystems and disrupting biodiversity. The soil in affected areas becomes dangerously contaminated, affecting agricultural productivity and posing health risks to the population.

– Several cases of illnesses, including cancer, have been attributed to radiation exposure from nuclear disasters and have been linked to an increased incidence of birth defects and other genetic abnormalities in affected regions.

– Additionally, the psychological impact of nuclear disasters extends to anxiety, depression and post-traumatic stress disorder among the affected population.

3.4.5. *Mitigation measures*

– *Regulatory frameworks*: stringent regulations are crucial to ensure the safe operation of nuclear facilities and minimize the risk of disasters.

– *Emergency preparedness and response*: this includes developing comprehensive emergency preparedness and response plans for nuclear accidents, including evacuation procedures, communication protocols, medical response capabilities and coordination mechanisms with local authorities, emergency services and relevant stakeholders. It also includes conducting regular drills, exercises and simulations to test emergency response procedures and enhance readiness to effectively manage nuclear emergencies.

– *Public awareness*: in this, comprehensive emergency response plans and drills are essential to mitigate the impact of potential nuclear incidents. It also includes ongoing training and education of nuclear facility personnel, emergency responders and the public on nuclear safety and response procedures, and clear and timely communication with the public to provide guidance, updates and reassurance during a nuclear disaster.

– *Safety protocols*: this includes implementation of rigorous safety protocols, including regular inspections, equipment maintenance and safety drills to minimize the risk of nuclear accidents.

– *Risk assessment*: this includes continuous assessment of potential risks and vulnerabilities at nuclear facilities, with proactive measures to address identified concerns.

– *Security measures*: this includes robust security measures to safeguard nuclear facilities from external threats, including unauthorized access and sabotage, theft or misuse of nuclear materials or facilities by malicious actors. Enhancing physical security, access controls, perimeter fencing, surveillance systems, cybersecurity measures and personnel screening to safeguard against security threats and ensure the integrity and security of nuclear facilities.

– *Radiation monitoring*: this includes establishment of systems for continuous monitoring of radiation levels to promptly detect and respond to any abnormal readings.

– *Evacuation and sheltering*: this includes organized evacuation of affected areas and provision of safe shelter for displaced individuals to minimize radiation exposure.

3.4.6. *Safety measures for nuclear hazards*

3.4.6.1. *Dos*

– Go indoors and stay inside.

– Turn on the TV or radio and listen to announcements from your local government.

– Close doors/windows.

– When consuming food or water, make sure it is covered.

– If you are outside, cover your body and face with a wet handkerchief.

– Return home, change/remove clothes. Have a complete wash and use fresh clothing.

– Adhere to the instructions provided by emergency authorities and follow evacuation orders, if necessary.

– Give local authorities your whole cooperation and follow their directions, whether it is for medication or evacuation.

– You also need to be aware about the risk posed by nuclear radiation.

– If advised, seek shelter in a sturdy building or underground to minimize exposure to radiation.

3.4.6.2. *Don'ts*

– Do not panic.

– Refrain from spreading unverified information or rumors that could cause unnecessary panic.

– Avoid going outside or remaining outside.

– Avoid food, drink or milk from outside sources; water from open wells or ponds; and exposed crops and vegetables.

– Avoid contact with surfaces or materials that may be contaminated with radioactive substances.

– Do not disobey any instruction of the district or civil defense authorities for the safety of your family and property.

– Do not enter areas that have been designated as restricted or hazardous due to the nuclear incident.

Communities and individuals can lessen the possible impact on public health and safety by enacting safety measures and adhering to these dos and don'ts when it comes to nuclear dangers.

Figure 3.2. *Chemical hazards. For a color version of this figure, see www.iste.co.uk/akbar/disastermanagement.zip*

3.5. Chemical disaster

3.5.1. *Meaning and concept*

Chemical disasters are the catastrophic events involving release of hazardous chemical substances, whether in liquid, gas, or solid form, that can cause significant risks to human health, environment and property. These disasters can occur due to accidents, industrial incidents, transportation mishaps or intentional acts. Promoting safety, safeguarding public health and reducing the negative effects that chemical disasters have on the environment and the economy all depend on an understanding of the causes, effects and preventive actions associated with them. These mishaps can happen in a variety of places, such as transportation hubs, storage facilities and industrial facilities; mishaps in chemical manufacturing plants, refineries, or storage facilities can lead to leaks, explosions or fires, causing the release of toxic substances. Toxic chemical releases can occur as a result of mishaps, malfunctioning machinery, mistakes made by people, or deliberate actions. Accidents during the transportation of hazardous chemicals via road, rail or waterways can result in spills or leaks, endangering nearby populations and ecosystems. In addition, deliberate acts of terrorism or sabotage aimed at chemical facilities or transportation routes can lead to catastrophic releases of hazardous materials. Chemical disasters comprise a broad spectrum of incidents involving leakage, spillage or contact with dangerous chemicals, which present substantial dangers to human well-being, the ecosystem and infrastructure. Chemical disasters can take many different forms, ranging from small-scale events that occur in industrial facilities to massive catastrophes that impact entire towns. These catastrophes could involve the discharge of harmful gases, detonations, conflagrations or tainted air, land or waterways. A variety of things can lead to chemical disasters, such as terrorist or purposeful acts of sabotage, equipment failure, human mistakes and natural disasters. Chemical catastrophes can differ in their intensity and duration, based on the kind and amount of chemicals involved, how close they are to populated regions and how well emergency response plans work. To reduce damage and safeguard public safety, chemical disasters must be mitigated with careful preparation, strong risk management techniques and coordinated response activities from a variety of stakeholders.

3.5.2. *Causes of chemical disasters*

Chemical accidents can result from a multitude of factors, most frequently from mistakes made by humans, malfunctions in technology or natural occurrences.

Chemical mishaps in industrial facilities are often caused by human error, which includes insufficient equipment maintenance, inappropriate handling or storage of dangerous chemicals, and disregard for safety procedures. Toxic materials may seep into the environment as a result of technological breakdowns, such as broken pipes or storage tank ruptures.

Chemical catastrophes can also be caused by natural disasters like earthquakes, floods or harsh weather, which can harm infrastructure or interfere with safety measures. Furthermore, deliberate acts of terrorism or sabotage directed toward chemical facilities have the potential to cause disastrous events with far-reaching effects. Comprehending the diverse origins of chemical accidents is imperative to institute pre-emptive actions, improve safety procedures and alleviate the hazards linked to the handling and preservation of dangerous substances.

Chemical disasters can arise from a complex interplay of various factors, each contributing to the likelihood and severity of the event. Firstly, inadequate safety measures and lax regulatory enforcement can significantly increase the risk of chemical accidents. Poor maintenance of equipment, outdated technology and insufficient training of personnel also amplify the chances of mishaps. Additionally, human error, whether in the form of improper handling of hazardous materials or miscommunication during operations, can escalate into catastrophic incidents.

Malfunction or failure of industrial equipment, such as storage tanks, pipelines, or processing units, can lead to chemical leaks, spills or releases. Errors in handling, transferring or processing hazardous chemicals, as well as lapses in safety protocols, can contribute to industrial accidents.

External factors such as natural disasters, sabotage or terrorist attacks further compound the vulnerability of chemical facilities. Moreover, population density around chemical plants can exacerbate the potential consequences of these hazards, especially if evacuation procedures are inadequate.

Finally, the lack of public awareness and community preparedness about chemical hazards can impede timely response efforts and exacerbate the overall impact of such disasters. Addressing these multifaceted factors requires a comprehensive approach that encompasses regulatory reform, improved safety protocols, enhanced training and heightened public awareness.

3.5.3. *Some of the worst chemical disasters*

1) *Texas City disaster (1947)*: the Texas City disaster occurred when a ship loaded with ammonium nitrate exploded in the port of Texas City, Texas. The explosion and subsequent fires killed hundreds of people and caused extensive damage to the surrounding area.

2) *Bhopal gas tragedy (1984)*: the Bhopal gas tragedy occurred in Bhopal, India, when a toxic gas (methyl isocyanate) leaked from pesticide plant storage tanks (owned by Union Carbide Corporation) due to a combination of factors, including inadequate safety measures, poor maintenance and human error. The incident resulted in thousands of immediate deaths and long-term health issues for those exposed. The Bhopal gas tragedy is one of the world's deadliest industrial disasters. Tens of thousands of others suffered long-term health effects, including respiratory

issues, eye problems and chronic illnesses. The impact also extended to subsequent generations due to the lingering effects of exposure.

3) *West Fertilizer plant explosion (2013)*: the explosion at the West Fertilizer Company plant in Texas involved the release of ammonium nitrate stored at the facility. The blast caused multiple casualties, destroyed nearby structures and raised concerns about the storage and handling of hazardous materials.

4) *BP Texas City refinery explosion (2005)*: an explosion at the BP Texas City refinery resulted in the release of flammable hydrocarbons and toxic substances. The incident caused multiple fatalities, injuries and extensive damage to the refinery.

5) *Sandoz chemical spill (1986)*: the Sandoz chemical spill in Switzerland resulted in the release of tons of toxic chemicals into the Rhine River, causing massive killing of fishes and other aquatic species and environmental damage across national borders. The incident prompted international cooperation to address transboundary environmental disasters and led to improvements in chemical storage and handling practices.

3.5.4. *Impact of chemical disasters*

Chemical disasters, characterized by the release of hazardous substances into the environment, can have devastating impacts on human health, ecosystems and socioeconomic stability. The release of toxic chemicals, whether through industrial accidents, transportation incidents or deliberate acts, can result in acute and chronic health effects, including respiratory problems, skin irritation, neurological disorders and cancer. Contamination of air, water, soil and food supplies with hazardous chemicals poses risks to human populations, wildlife and ecosystems, leading to ecosystem degradation, loss of biodiversity and long-term environmental damage.

Chemical disasters can also disrupt communities, economies and infrastructure, causing property damage, displacement of populations, loss of livelihoods and economic losses. Exposure to toxic chemicals can cause immediate health effects, such as respiratory problems, burns and chemical poisoning. Long-term exposure may lead to chronic health conditions. In addition, chemical disasters can result in significant economic losses due to clean-up and recovery efforts, healthcare costs, property damage and impact on local industries and businesses. Cleanup and remediation efforts following chemical disasters require significant resources, expertise and coordination to mitigate environmental contamination and restore affected areas to safe and habitable conditions. Effective prevention, preparedness and response measures are essential for minimizing the risks of chemical disasters and protecting public health, safety and the environment from the adverse impacts of hazardous chemical releases.

3.5.5. *Prevention and mitigation*

– *Regulatory measures*: this step includes implementing and enforcing strict safety regulations, including the proper handling, storage and transportation of hazardous chemicals.

– *Risk assessment*: personnel should conduct appropriate risk assessments and develop emergency plans and training to mitigate the effects of potential chemical hazards.

– *Public awareness and community engagement*: this includes educating people and communities about chemical hazards/disasters, and emergency response procedures, and fostering public participation in disaster preparedness and response efforts.

– *Containment and cleanup*: this includes establishing protocols for containing and cleaning up chemical spills to prevent further contamination and minimize environmental impact.

– *Emergency response plan and preparedness*: this includes developing and regularly reviewing emergency response plans, including evacuation procedures and communications plans. Additionally, it collaborates with local authorities and communities to develop and implement response plans, conduct drills raise awareness, and prepare for potential chemical hazards.

3.5.5.1. *Safety tips*

Safety tips for chemical disasters are important to minimize risks and protect individuals in the event of an incident involving hazardous chemicals. The following are some essential safety tips to reduce the harmful effects of chemical hazards:

– *Stay informed*: be aware of potential chemical hazards in your area and stay informed about emergency procedures and evacuation routes.

– *Follow instructions*: if authorities issue warnings or evacuation orders, follow them promptly and cooperate with emergency responders.

– *Evacuate safely*: leave the affected area immediately if instructed to do so. Follow designated evacuation routes and avoid shortcuts through potentially hazardous areas.

– *Shelter in place*: if evacuation is not possible or advised, seek shelter indoors in a sturdy building. Close all windows and doors, and seal gaps with duct tape or wet towels to prevent chemical vapors from entering.

– *Protect yourself*: wear protective clothing, such as long sleeves, pants, gloves and goggles, to minimize skin contact and exposure to hazardous chemicals.

– *Avoid contaminated areas*: stay away from areas where chemicals have spilled or leaked, as well as downwind of the incident to avoid exposure to toxic fumes.

– *Seek medical attention*: if you come into contact with hazardous chemicals or experience symptoms of exposure (e.g. difficulty breathing, skin irritation, nausea), seek medical attention immediately.

– *Prevent spread*: do not touch or handle any spilt chemicals and avoid spreading contaminated materials to unaffected areas.

– *Stay informed*: listen to emergency alerts and updates from local authorities via radio, television or official social media channels for instructions and updates.

3.6. Technological disasters

3.6.1. *Meaning and concept*

Technical hazards/disasters refer to incidents or problems caused by malfunctions, human error, failure or misuse of industrial systems and equipment. These hazards can have widespread and severe consequences for human safety, the environment and public welfare. The economy, human health and the environment are all at risk from these disasters. These types of disasters can also include transportation accidents involving aircraft, railroads or ships; industrial accidents such as chemical spills or nuclear explosions; and cyberattacks that cause severe infrastructure damage.

Technological mishaps frequently result from a confluence of events, such as defective machinery, insufficient safety precautions and human error. In addition, sabotage, terrorism or intentional misuse of technology can also lead to technological disasters and can have severe consequences for society. These catastrophes can have far-reaching effects, including fatalities, community uprooting, environmental damage and high financial expenses. Understanding the origins and effects of technological disasters is crucial as society depends more and more on intricate technological systems; it is critical to comprehend the origins and effects of technology disasters in order to reduce possible effects or risks and improve general safety and resilience. Technological disasters are most common in nations like Iran, where the most serious technological risks are transport accidents (road, airplane and train). These accidents cause a great deal of death and injury, as well as significant financial losses and other negative effects on the environment, human health and property. These tragedies are frequently caused by both human error and technological flaws, so it is critical to have emergency response plans in place, continually monitor and control technology, and train personnel on new technologies to minimize errors. In these nations, the rapid growth of transportation networks and urbanization are two more factors contributing to the frequency of technology disasters. The nature of technology failure can vary widely, and it often results from a combination of factors, including design flaws, inadequate maintenance and unforeseen circumstances.

3.6.2. *Causes of technological disasters*

These disasters can be caused by various factors, including design flaws, equipment malfunction, human error, inadequate maintenance or intentional actions.

– *Critical infrastructure failure*: technological disasters often involve the failure of critical infrastructure, such as power grids, transportation systems, communication networks or industrial facilities. The disruption of these systems can have cascading effects on society.

– *Human error*: technological disasters may result from errors in human–machine interaction, including mistakes made during the design, operation or maintenance of technology. Miscommunication, inadequate training, negligence or lapses in human judgment can contribute to disasters.

– *Equipment failure*: malfunctions or breakdowns in technological equipment, machinery or components can result in disasters. This can be due to manufacturing defects, wear and tear, or inadequate maintenance.

– *Unintended consequences*: some technological disasters occur due to unintended consequences of technological advancements. For instance, the release of hazardous substances during industrial processes or the unintended side effects of new technologies can lead to disasters.

– *Design flaws*: inadequate or flawed design of technological systems can lead to vulnerabilities and failures. Design flaws may become apparent over time or under specific conditions, contributing to disasters.

– *Cyberattacks*: malicious activities, such as cyberattacks, hacking or the introduction of malware, can compromise the security and functionality of technological systems. Cybersecurity breaches can lead to data loss, system failures or disruption of critical services.

– *Industrial accidents*: technological disasters may result from accidents in industrial facilities, including chemical spills, explosions or nuclear incidents. Poor safety practices, inadequate risk assessment or equipment failure can contribute to such accidents.

3.6.3. *Impact of technological disasters*

Technological disasters can affect human communities, the environment and economies in a wide range of complex ways. The ramifications of a disaster can be multifaceted and contingent upon its characteristics, the technologies entailed and the established response protocols. The following are some typical effects of technology mishaps on human societies, the environment and economies. The consequences are often complex and can vary depending on the nature of the disaster, the technologies involved.

– *Loss of life and long-term health effects*: technological disasters can result in immediate injuries and fatalities, especially if they involve explosions, chemical releases or other catastrophic events. Exposure to hazardous materials and pollutants during technological disasters can lead to long-term health issues, including respiratory problems, cancers and other chronic conditions.

– *Environmental degradation*: accidents involving the release of hazardous chemicals, oil spills or radioactive materials can cause severe environmental pollution, impacting ecosystems, water bodies and soil quality. The release of pollutants into the air or water can have long-lasting effects on the environment, affecting flora and fauna and compromising biodiversity. The long-term environmental consequences of technological disasters can persist in ecosystems, affecting biodiversity, water quality and soil fertility.

– *Infrastructural and property damage*: explosions, fires or structural failures associated with technological disasters can lead to significant damage to buildings, roads, bridges and other critical infrastructure. The destruction of industrial facilities and infrastructure can result in substantial economic losses for both businesses and affected communities. Disruption to businesses and agriculture can result in the loss of livelihoods for individuals and communities.

– *Social and psychological impact*: technological disasters can cause emotional distress, trauma and anxiety within affected communities, impacting mental health and well-being. Displacement, loss of loved ones and economic hardships can lead to social disruptions, including increased vulnerability.

– *Disruption of critical services*: technological disasters can disrupt essential services such as water supply, power and communication networks, making recovery efforts more challenging. Disasters may overwhelm healthcare systems, hindering the ability to provide medical care and respond to health emergencies.

– *Contamination of food and water supply*: contamination of soil and water can affect crops, reducing agricultural productivity and posing risks to food safety. Pollutants released during disasters can enter the food chain, affecting both plant and animal sources of food.

3.6.4. *Mitigation measures of technological disasters*

Mitigation measures for technological disasters aim to reduce the risk, impact and severity of disasters related to technology failures or accidents.

These measures involve a combination of planning, regulation, technology improvements and community preparedness. The following are some key mitigation measures:

– *Risk assessment and planning*: conduct thorough risk assessments to identify potential hazards and vulnerabilities associated with technology systems and infrastructure. Develop comprehensive emergency response plans that outline actions to be taken in the event of a technological disaster, including evacuation procedures, communication strategies and resource allocation.

– *Regulatory oversight and compliance*: implement and enforce strict regulations and standards for industries involving hazardous technologies, such as nuclear power, chemical manufacturing and oil drilling. Regularly inspect and monitor technological facilities to ensure compliance with safety standards and identify potential risks.

– *Technological improvements*: integrate advanced safety features and fail-safe mechanisms into technological systems to minimize the risk of accidents. Invest in research and development to advance technologies that can improve safety, reduce environmental impact and enhance disaster resilience.

– *Public education, awareness and training programs*: educate the public about potential technological hazards, emergency procedures and the importance of preparedness through public awareness campaigns. Provide specialized training for workers in high-risk industries to enhance their knowledge of safety protocols and emergency response. Conduct regular drills and exercises involving local communities, emergency responders and relevant stakeholders to test and improve preparedness. Develop and communicate evacuation plans for communities residing near high-risk technological facilities.

– *Early warning and monitoring systems*: implement advanced monitoring systems to detect early signs of technological failures, enabling rapid response and intervention. Also, establish robust communication networks for timely dissemination of warnings and information to the public and relevant authorities.

– *Environmental impact assessment*: conduct thorough environmental impact assessments for technological projects to evaluate potential consequences and identify measures to minimize adverse effects. Develop and implement mitigation strategies to reduce the environmental impact of technological disasters, such as oil spills or chemical releases.

– *Infrastructure resilience*: build redundancy into critical infrastructure systems to ensure continued functionality in the event of component failures. Implement resilient design principles in the construction of technological facilities to withstand natural disasters, such as earthquakes or hurricanes.

– *International collaboration*: foster international collaboration for information sharing, best practices and joint research to enhance global resilience to technological disasters. In addition, develop comprehensive plans for post-disaster recovery and reconstruction.

3.7. Questions

1) Discuss in detail transportation disasters (road accidents), their causes and mitigation measures.

2) Explain in detail chemical disasters, their causes and preventive methods for mitigation.

3) Write a short note on the Bhopal Gas tragedy.

4) What are anthropogenic disasters? Give three examples.

5) What are nuclear disasters? Discuss their effects in detail.

6) Explain briefly the concept of biological hazards with suitable examples.

7) What are urban fires, and how do they differ from wildfires?

8) Describe the main causes and contributing factors of urban fires in densely populated areas.

9) Explore the factors that contribute to the occurrence of chemical disasters, such as industrial accidents, transportation incidents, chemical warfare and terrorism.

10) What are some safety tips for nuclear disasters?

11) How do human errors and technical failures contribute to technology disasters?

12) How can organizations prepare for and respond to technology disasters, such as system failures, to minimize their impact?

13) Discuss Chernobyl and Fukushima accidents as case studies.

14) What lessons have been learned from past nuclear disasters to improve safety and preparedness measures?

15) What measures can be taken to mitigate the impact of a nuclear disaster?

16) What are some common factors contributing to transport disasters?

17) What measures can be taken to mitigate the risks and consequences of transport disasters?

18) Investigate the societal and economic implications of human-induced hazards, examining how communities and industries respond to and recover from such disasters.

19) What are the key components of the National Disaster Management Authority (NDMA) guidelines about biohazards?

20) What are some examples of biological disasters, and what factors contribute to their occurrence?

21) What are some effective strategies for engaging urban populations in fire safety initiatives?

22) How do radiation disasters affect human health, the environment and infrastructure?

23) What are the socioeconomic impacts of human-induced disasters on communities and regions affected by them?

24) Discuss the impact of technology disasters on environment.

25) What measures can be taken to mitigate the risks and consequences of technology disasters?

26) What are the dos and don’ts of chemical disasters?

27) What are some factors contributing to a nuclear disaster?

28) What are the primary causes of urban fires, and how do they differ across residential, commercial and industrial settings?

29) What are the primary factors contributing to human-induced environmental disasters?

30) What are the key challenges faced by urban fire departments in densely populated areas?

3.8. Multiple-choice questions (MCQs)

1) A disease that becomes unusually widespread and an event global in its reach is referred to as:

a) Epidemic

b) **Pandemic**

c) Spanish flu

d) Hyper endemic

2) What is the primary cause of an industrial explosion?

a) Earthquake

b) **Human error**

c) Volcanic activity

d) Tornado

3) What human-induced disaster involves the intentional release of harmful chemicals into the environment?

a) Oil spill

b) Industrial accident

c) Nuclear meltdown

d) **Biological warfare**

4) What is the primary cause of an oil spill in marine environments?

a) Volcanic eruptions

b) **Human error during transportation**

c) Earthquakes

d) Industrial accidents

5) Which human-induced disaster is associated with the collapse of large structures like bridges or buildings?

a) Industrial accident

b) Nuclear meltdown

c) **Structural failure**

d) Oil spill

6) What is a common cause of water pollution from industrial activities?

a) Agricultural runoff

b) **Oil spills**

c) Deforestation

d) Pesticide contamination

7) What human-induced disaster is characterized by the release of radioactive materials into the environment?

a) **Nuclear meltdown**

b) Industrial explosion

c) Oil spill

d) Hazardous waste disposal

8) What is the primary cause of a nuclear meltdown in a power plant?

a) Earthquake

b) **Human error**

c) Volcanic eruption

d) Tsunami

9) What human-induced disaster involves the sudden and widespread failure of the power grid?

a) **Cyberattacks**

b) Industrial explosion

c) Oil spill

d) Nuclear meltdown

10) What human activity is a major contributor to the depletion of groundwater resources?

a) Deforestation

b) **Over-extraction for agriculture**

c) Industrial emissions

d) Nuclear accidents

11) Which of the following is a consequence of improper waste disposal in oceans?

a) Acid rain

b) **Coral bleaching**

c) Tornado

d) Soil erosion

12) What human-induced disaster involves the contamination of air by harmful substances?

a) Industrial accident

b) Oil spill

c) **Air pollution**

d) Nuclear meltdown

13) What is the significant environmental impact of uncontrolled deforestation?

a) **Soil erosion**

b) Air pollution

c) Oil spills

d) Nuclear accidents

14) Which of the following human-induced disasters is associated with the release of harmful substances into the soil?

a) Nuclear meltdown

b) **Industrial contamination**

c) Oil spill

d) Biological warfare

15) Which of the following is a consequence of nuclear weapons testing?

a) Greenhouse effect

b) **Radioactive contamination**

c) Landslides

d) Acid rain

16) Which of the following human-induced disasters involves the collapse of dams or reservoirs?

a) Industrial accident

b) Nuclear meltdown

c) **Dam failure**

d) Oil spill

17) What is a major cause of chemical contamination in rivers and lakes?

a) Industrial accidents

b) **Oil spills**

c) Deforestation

d) Volcanic eruptions

18) Mass killing diseases can be referred to as:

a) **Biological disaster**

b) Industrial disaster

c) War disaster

d) Flood disaster

19) "Influenza" can also be called:

a) Groin

b) **Flu**

c) Both a and b

d) None of the above

20) What is the main cause of cholera?

a) **Poor sanitation**

b) Floods

c) Air pollution

d) Poor nutrition

21) An example of anthropogenic hazards is:

a) **Chemical explosion**

b) Earthquake

c) Flood

d) Wildfire

22) The human factors which cause floods are:

a) **Deforestation**

b) Heavy rainfall

c) Cloud burst

d) Large catchment area

23) The Bhopal gas tragedy is an example of:

a) **Industrial disasters**

b) Natural disasters

c) Nuclear disasters

d) None

24) Pandemic disease is defined as:

a) **Outbreak of a disease in an international scale**

b) Outbreak of a disease beyond the area of a disease

c) Congestion in urban areas

d) Congestion in rural areas

25) Industrial hazards come under the category of:

a) Natural hazards

b) **Human-induced hazards**

c) Meteorological hazard

d) Wildfire hazard

26) Pandemic diseases include:

a) Cholera

b) Smallpox

c) Tuberculosis

d) **All of the above**

27) Disasters at Chernobyl (USSR), Three Mile Island (USA) and Fukushima (JAPAN) are all related to:

a) Disasters caused by earthquakes

b) Disasters caused by tsunamis

c) **Nuclear disasters**

d) Epidemic disasters

28) The disease that swept a large part of Asia, Europe and Africa in the 14th century was:

a) **Plague**

b) Smallpox

c) Tuberculosis

d) Influenza

29) Zika virus is related to:

a) Dengue

b) Yellow fever

c) Japanese encephalitis

d) **All the above**

30) The 2001 terrorist attacks in America are referred to as:

a) **9/11**

b) 11/9

c) 10/11

d) 11/01

4

Pollutions

4.1. Meaning and concept

Pollution is defined as the presence of hazardous and artificial substances that affect the ecosystem and provide health risks to both people and animals. This phenomenon occurs when harmful unnatural ingredients replace or contaminate natural ingredients, potentially throwing the ecosystem out of balance and posing numerous health risks to both humans and animals.

A pollutant is waste material that pollutes air, water or soil and is the cause of pollution. A pollutant's intensity is determined by three factors: its chemical composition, concentration and persistence. Certain contaminants are biodegradable, meaning they will not stay in the environment for very long. There are two types of pollutants: primary and secondary. Primary pollutants are emitted directly into the atmosphere from sources such as SO_2, NOx, chlorofluorocarbons (CFCs), carbon monoxide (CO), volatile organic compounds (VOCs) and particulate matter (PM). These pollutants have a significant impact on the environment and human health. A secondary pollutant is not directly emitted from sources. Instead, they form in the atmosphere through chemical reactions among pollutants and other atmospheric compounds. These pollutants include ozone (O_3) (ground-level ozone), photochemical smog, H_2SO_4 and HNO_3.

For a color version of all figures in this chapter, see www.iste.co.uk/akbar/disastermanagement.zip.

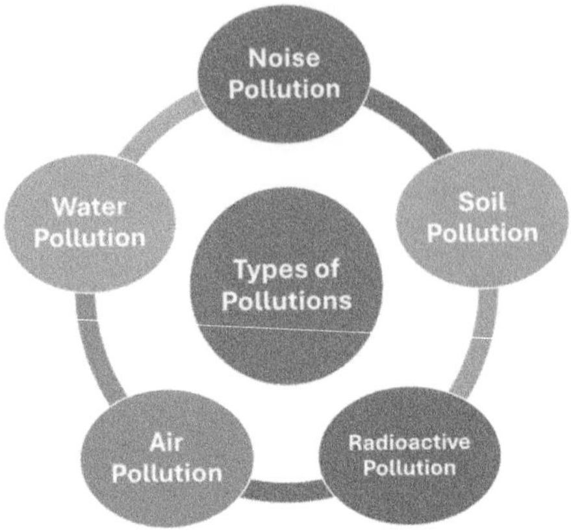

Figure 4.1. *Types of pollution*

4.2. Air pollution

4.2.1. *Meaning and concept*

Air pollution refers to the contamination of air due to the presence of harmful substances in the Earth's atmosphere, which can have adverse effects on human health, environment and climate. These pollutants can be natural or human made and come from various sources, such as industrial activities, transportation and agricultural practices. Some of the most prominent air pollutants are CO, O_3 and PM. Pollutants such as PM_{10}, $PM_{2.5}$, nitrogen dioxide, sulphur dioxide (SO_2) and VOCs cause health problems, especially in children.

Figure 4.2. *Toxic gases from industries polluting air*

4.2.2. Sources of air pollution

Air pollution has various causes, and it often results from a combination of natural processes and human activities. The primary sources of air pollution include the following:

1) *Combustion of fossil fuels*: the burning of fossil fuels such as coal, oil and natural gas in power plants, industrial facilities and vehicles is a major contributor to air pollution. This combustion releases pollutants such as carbon dioxide (CO_2), nitrogen oxides (NOx), SO_2, PM and VOCs.

2) *Transportation*: emissions from vehicles, including cars, trucks, buses, and motorcycles, release pollutants such as carbon monoxide (CO), nitrogen oxides (NOx), PM and VOCs. Traffic congestion and inefficient transportation systems can exacerbate these emissions.

3) *Industrial activities*: manufacturing processes, chemical production and other industrial activities release a significant amount of pollutants into the air. Common emissions include PM, SO_2, nitrogen oxides, VOCs and various industrial chemicals.

4) *Agricultural practices*: agricultural activities contribute to air pollution through the release of ammonia (NH_3) from fertilizers, methane (CH_4) from livestock and various volatile organic compounds. Pesticides and herbicides used in agriculture can also release harmful chemicals into the air.

5) *Waste management*: improper disposal and management of solid waste, including open burning of waste, can release pollutants into the air. Incineration of waste materials can also produce harmful emissions.

6) *Deforestation and land use changes:* changes in land use, such as deforestation and urbanization, can affect air quality. Forest fires, whether natural or human induced, release large amounts of PM and other pollutants into the atmosphere.

7) *Construction activities*: construction sites can generate dust and release pollutants into the air due to activities such as excavation, demolition and the use of heavy machinery.

8) *Natural sources*: certain natural processes contribute to air pollution. For example, wildfires release large amounts of PM and carbon monoxide. Volcanic eruptions can release SO_2 and other gases.

9) *Indoor sources*: indoor activities, such as cooking with solid fuels, smoking, and using certain household products, can contribute to indoor air pollution. This can have direct implications for respiratory health.

10) *Power generation*: besides fossil fuel combustion, power generation from sources like biomass burning and waste incineration can release pollutants into the air.

4.2.3. *Effects of air pollution*

The effects of air pollution are severe and pervasive. In humans, it can result in respiratory disorders, heart troubles and other health concerns. In addition, it can destroy crops, disrupt ecosystems and exacerbate climate change. Regulatory actions, technological developments aimed at lowering emissions and public awareness efforts are all part of the fight against air pollution. Some of the key consequences of air pollution include the following:

1) *Respiratory and cardiovascular health issues*: exposure to air pollutants, especially PM, nitrogen dioxide (NO_2), SO_2, and ozone (O_3), can lead to respiratory problems such as asthma, bronchitis and other chronic respiratory diseases. Long-term exposure is also associated with cardiovascular diseases.

2) *Increased mortality rates*: high levels of air pollution have been linked to premature deaths. Health studies have shown a correlation between air pollution and increased mortality rates, particularly due to respiratory and cardiovascular diseases.

3) *Impaired lung development and cancer in children*: children exposed to air pollution may experience impaired lung development, leading to long-term respiratory issues. Asthma in children is often aggravated by exposure to pollutants. Certain air pollutants, such as benzene, formaldehyde and polycyclic aromatic hydrocarbons (PAHs), are known carcinogens. Prolonged exposure to these substances increases the risk of developing cancer, particularly lung cancer.

4) *Ecosystem damage*: air pollution can harm ecosystems by affecting soil quality, water bodies and vegetation. Acid rain, resulting from SO_2 and nitrogen oxide emissions, can damage forests, lakes and soil, impacting both terrestrial and aquatic ecosystems. In addition, ozone and PM can negatively affect crops, reducing yields and impacting food quality. Ground-level ozone

can interfere with photosynthesis and damage plant tissues, leading to economic losses for farmers.

5) *Climate change*: certain air pollutants, such as carbon dioxide (CO_2) and methane (CH_4), are greenhouse gases that contribute to climate change. The warming of the Earth's atmosphere has wide-ranging effects, including altered weather patterns, rising sea levels and disruptions to ecosystems.

6) *Visibility reduction*: PM and other pollutants can reduce visibility, posing safety risks for transportation and impacting the aesthetic value of landscapes and tourist destinations. Acid rain is a significant environmental issue that has garnered attention due to its detrimental effects on the environment.

4.2.4. Control measures of air pollution

The following are some crucial countermeasures for air pollution:

1) *Emission standards and regulations*: governments can establish and enforce strict emission standards for industries, power plants and vehicles. These regulations limit the amount of pollutants that can be released into the air, encouraging the use of cleaner technologies and processes. In addition, the introduction of the air quality index, Bharat Stage Emission Standard and Air Act 1981, the National Air Quality Monitoring Programme (NAQMP), etc., are some government initiatives to control air pollution.

2) *Transition to clean energy*: promoting the use of suitable, clean and renewable energy sources, such as solar, wind and hydropower, helps reduce the reliance on fossil fuels to decrease emissions from power generation. Implementing standards for cleaner fuels, such as low-sulphur fuels, reduces the emissions of SO_2 and other pollutants from vehicles and industrial processes.

3) *Public transportation and sustainable mobility*: encouraging the use of public transportation, walking, cycling and electric vehicles can reduce emissions from the transportation sector. Investment in sustainable urban planning and infrastructure is crucial for promoting cleaner mobility options.

4) *Waste management practices*: proper waste disposal and recycling methods can help reduce the release of pollutants from burning waste. Implementing waste-to-energy technologies can also be considered to generate energy from waste materials.

5) *Afforestation and urban greenery*: planting more and more trees and creating green spaces in urban areas can help absorb pollutants, improve air quality and provide multiple environmental benefits. Trees act as natural filters, capturing PM and absorbing certain gases.

6) *Monitoring and reporting*: establishing air quality monitoring systems helps identify pollution sources and assesses the effectiveness of mitigation measures. Real-time data can inform policymakers and the public about air quality issues.

7) *Education and awareness*: increasing public awareness about the sources and impacts of air pollution can minimize its effects. Educational initiatives can encourage behaviors that lessen individual and community's contribution to air pollution.

4.3. Water pollution

4.3.1. *Meaning and concept*

Water pollution refers to the contamination of water bodies, such as rivers, lakes, seas, oceans, aquifers, groundwater and other aquatic systems, by the introduction of harmful substances that are discharged either directly or indirectly into bodies of water without sufficient treatment. These substances can be physical, chemical or biological and have adverse effects on the water quality, the ecosystem and human health. Water pollution can result from various human activities and natural processes. Numerous factors contribute to this contamination, such as oil spills, inappropriate waste disposal, runoff from agriculture and industrial discharge. It is a critical environmental concern and has serious negative consequences on human health, aquatic habitats, tourism and fishing, as well as other economic endeavours. Coordinated actions are needed to address water contamination, such as strict laws, environmentally friendly waste management, infrastructure spending and public awareness initiatives. We can guarantee the maintenance of clean and safe water sources for present and future generations by giving mitigation of water pollution priority. Water pollution is a pressing environmental issue with far-reaching consequences for ecosystems, human health and socioeconomic development. At its core, water pollution encompasses the contamination of water bodies such as rivers, lakes, oceans and groundwater by harmful substances or pollutants. These pollutants can originate from various sources, including industrial discharges, agricultural runoff, untreated sewage and improper waste disposal. The impacts of water pollution are manifold, affecting aquatic life through habitat destruction, species displacement and the accumulation of toxins in the food chain. Moreover, contaminated water poses significant risks to human health, leading to waterborne diseases, reproductive disorders and even cancer in some cases. Additionally, water pollution exacerbates water scarcity issues by rendering vast quantities of water unfit for consumption or agricultural use. Addressing water pollution requires a multifaceted approach involving strict regulations, sustainable resource management practices, investment in wastewater treatment infrastructure and public education initiatives to foster a culture of environmental stewardship. Water pollution has a significant impact on the ecosystem and human well-being. It can harm aquatic life, disrupt food chains, degrade water quality for drinking and recreational purposes and pose health risks to humans and animals. By safeguarding water quality, we can ensure the health and well-being of both ecosystems and human populations. Mitigating water pollution requires a combination of regulatory actions, technological solutions and changes in behavior. Efforts to mitigate water pollution involve implementing and enforcing environmental regulations, improving wastewater treatment infrastructure, adopting sustainable agricultural practices and promoting public awareness about the importance of water conservation and pollution prevention.

Figure 4.3. *Polluted water body*

4.3.2. *Causes of water pollution*

– *Industrial discharges*: effluents from industrial processes can contribute to water contamination. Airborne pollutants, including heavy metals and industrial chemicals, can settle on water surfaces through precipitation (acid rain) or atmospheric deposition. These pollutants may then enter water bodies, affecting water quality. Factories and industrial facilities often release pollutants such as heavy metals, chemicals and toxins into nearby water bodies.

– *Agricultural runoff*: the use of fertilizers, pesticides and herbicides in agriculture can lead to the runoff of these chemicals into nearby rivers and streams. This agricultural runoff may contain nutrients, like nitrogen and phosphorus, which can contribute to water pollution and algal blooms.

– *Wastewater and sewage disposal*: improper disposal of domestic sewage and untreated wastewater from households, sewage treatment plants and inadequate sanitation systems can introduce bacteria, pathogens and pollutants into water bodies.

– *Chemical and oil spills*: accidental releases of hazardous chemicals or spills during transportation can result in the contamination of nearby waterbodies, posing risks to aquatic ecosystems and drinking water supplies and causing long-lasting damage to marine environments.

– *Mining activities*: mining operations can release heavy metals and toxic substances into water bodies, contaminating them. Acid mine drainage is a specific problem where acidic water containing metals is released from mining sites.

– *Urban runoff*: urban areas with impervious surfaces, such as roads and pavements, can lead to increased surface runoff during rainfall. This runoff may carry pollutants such as oil, heavy metals and litter into storm water drains and ultimately into water bodies.

– *Plastic pollution*: improper disposal and inadequate waste management contribute to the accumulation of plastic debris in oceans and waterways. Microplastics, small plastic particles, can also contaminate water and pose risks to aquatic life.

4.3.3. *Effects of water pollution*

Effective water pollution prevention and management strategies involve regulatory measures, improved wastewater treatment, sustainable agricultural practices and public awareness campaigns. The following are some of the key impacts of water pollution:

1) *Loss of biodiversity*: water pollution can lead to the decline and loss of aquatic biodiversity. Pollutants, such as chemicals and toxins, can directly harm fish, amphibians and other aquatic organisms. Changes in water quality and habitat degradation can also disrupt ecosystems and lead to the loss of species.

2) *Harm to aquatic life*: pollutants like heavy metals, pesticides and industrial chemicals can be toxic to aquatic organisms. They may cause deformities, reproductive issues and even death in fish, amphibians and other aquatic species.

3) *Disruption of food chains*: water pollution can disrupt food chains and aquatic ecosystems. Contaminated organisms may pass toxins up the food chain, affecting predators and potentially impacting human health if the polluted organisms are part of the human diet.

4) *Contamination of water*: water pollution poses a significant threat to drinking water supplies. Contaminants such as bacteria, viruses, pesticides, heavy metals and industrial chemicals can make surface water unsafe for human consumption. Excessive nutrient runoff, often from agricultural activities, can lead to the overgrowth of algae in water bodies. This can result in harmful algal blooms (HABs), which produce toxins harmful to aquatic life and can lead to "dead zones" with low oxygen levels, negatively impacting fish and other organisms. In addition, pollutants can infiltrate the soil and contaminate groundwater, which serves as a source of drinking water for many communities, posing long-term risks to human health, which leads to waterborne diseases and other health issues.

5) *Economic impact*: water pollution can have economic repercussions, particularly in industries dependent on clean water, such as fisheries and tourism. Contaminated water bodies may lead to decreased fishery yields, loss of recreational opportunities and damage to local economies.

6) *Loss of recreational opportunities*: polluted water bodies are often unsuitable for recreational activities such as swimming, boating and fishing.

This can impact the quality of life for communities and reduce the economic value of recreational areas.

7) *Agricultural impact*: water pollution from agricultural runoff can harm crops, reduce soil fertility and affect agricultural productivity. Excessive nutrient runoff can lead to soil erosion and nutrient imbalances, impacting the sustainability of farming practices.

8) *Human health risks*: consuming or coming into contact with polluted water can pose significant health risks to humans. Waterborne diseases, such as cholera and dysentery, can spread through contaminated water sources. Additionally, exposure to certain pollutants may lead to chronic health problems, including cancer and developmental disorders.

4.3.4. *Control measures of water pollution*

Mitigating water pollution requires a combination of actions and technological solutions. Efforts to mitigate water pollution involve implementing and enforcing environmental regulations, improving wastewater treatment infrastructure, adopting sustainable agricultural practices, and promoting public awareness about the importance of water conservation and pollution prevention. The following are some key mitigation measures to address and prevent water pollution:

1) *Regulatory measures*: establish and enforce stringent water quality standards and regulations to limit the discharge of pollutants from industrial, agricultural and municipal sources. Implement and enforce effluent standards for industries to ensure that wastewater discharges meet specific quality criteria before entering water bodies.

2) *Wastewater and stormwater management*: upgrade and maintain wastewater treatment plants to effectively remove pollutants before discharging treated water into water bodies. Encourage the use of advanced treatment technologies to address emerging contaminants and improve the overall efficiency of wastewater treatment processes. Additionally, implement proper waste disposal and recycling programs to prevent hazardous materials from entering water bodies. Educate the public about the importance of proper waste management and discourage the illegal dumping of waste near the water bodies. Implement stormwater management practices to reduce runoff from urban areas, preventing the transport of pollutants into water bodies.

3) *Green infrastructure*: construct green infrastructure, such as permeable pavements and green roofs, to absorb and filter storm water before it reaches water bodies.

4) *Agricultural best practices*: promote and incentivize the adoption of sustainable agricultural practices that minimize the use of fertilizers, pesticides and herbicides. Encourage the use of cover crops, buffer zones and conservation tillage to reduce soil erosion and nutrient runoff from agricultural lands.

5) *Land use planning*: implement effective land use planning to prevent inappropriate development in sensitive areas, such as wetlands and watersheds. Establish and enforce zoning regulations to minimize pollution risks associated with urban and industrial development.

6) *Buffer zones and riparian restoration*: establish buffer zones along water bodies, particularly in agricultural areas, to filter pollutants before they reach the water. Promote riparian restoration projects to enhance the health of the zones along rivers and streams, preventing sedimentation and providing habitat for aquatic life.

7) *Public awareness and education*: generate mass social awareness about the impacts of water pollution and the importance of water conservation. Educate communities about proper disposal practices, water-friendly behaviors and the consequences of pollution on ecosystems and human health.

4.4. Soil pollution

4.4.1. *Meaning and concept*

Soil pollution refers to the contamination of soil with harmful substances, caused by the accumulation of toxic elements in the soil. This hazard poses a significant threat to human health, ecosystems and agricultural productivity. Contamination of soil with pollutants such as heavy metals, pesticides, industrial chemicals and hazardous waste can result from various human activities, including industrial processes, agricultural practices, mining operations, improper waste disposal and urbanization. Soil pollution can have detrimental effects on soil fertility, water quality and biodiversity, leading to reduced crop yields, loss of wildlife habitat and degradation of ecosystem services. In addition, contaminated soil can be dangerous to human health through the ingestion of contaminated food, exposure to contaminated water sources and inhalation of airborne pollutants. Addressing soil pollution involves comprehensive solutions that include remediation techniques, sustainable land management practices, pollution prevention measures and regulatory frameworks. Soil pollution is a pervasive environmental concern that arises from the accumulation of contaminants and pollutants in the Earth's soil layers. Chemical substances like heavy metals, pesticides, petroleum hydrocarbons and industrial by-products can seep into the soil, disrupting its natural composition and fertility. Soil pollution not only degrades soil quality but also poses serious threats to ecosystems, agricultural productivity and human health. Contaminated soil can harm plant growth, reduce crop yields and contaminate food supplies, leading to health problems when consumed. Furthermore, pollutants in the soil can leach into groundwater, contaminating water sources and posing risks to aquatic life and human populations. Addressing soil pollution requires comprehensive strategies, including remediation techniques, sustainable land management practices, regulatory measures to control pollutant discharge and public awareness campaigns to promote responsible waste management and land use practices. By maintaining soil quality, we can protect the environment, ensure food security and promote sustainable development for future generations. The build-up of toxic materials in the soil causes soil pollution, which has a profound effect on agriculture, human health, ecosystems and the environment. Pollutants like heavy metals, pesticides, industrial chemicals, and waste products erode soil structure, disturb soil fertility and impede plant growth. Contaminated soil can result in lower crop yields and compromised food safety, which in turn has an impact on agricultural production and food security. Ingestion of polluted food or water, absorption of airborne pollutants, or direct contact with contaminated soil are the other ways that soil contamination endangers human health. Furthermore, soil pollution damages soil-dwelling creatures and upsets ecological equilibrium, both of which have a negative effect on biodiversity. Polluted soil remediation is expensive and frequently difficult, requiring sustained work and investments in sustainable soil management techniques. Safeguarding ecosystems, advancing sustainable agriculture and preserving human health all depend on addressing soil pollution, underscoring the pressing need for efficient pollution prevention and repair techniques. Approximately 19.4% of arable land in China is polluted due to various activities such as smelting, mining, illegal sewage and waste discharge, as well as misuse and overuse of pesticides and fertilizers. This has led to significant heavy metal accumulation in the soil, particularly arsenic contamination, posing a threat to human health.

Figure 4.4. *Contaminated soil*

4.4.2. Causes of soil pollution

– *Industrial activities*: chemical spills, improper disposal of industrial waste, leakage from storage tanks and release of pollutants from factories can contaminate the soil by introducing toxic substances into it.

– *Agricultural activities*: overuse of pesticides, herbicides and highly toxic fertilizers can lead to soil pollution and contamination, especially when used inappropriately.

– *Improper waste disposal*: open dumping of household waste, municipal solid waste, sewage sludge construction debris and electronic waste can introduce harmful substances into the soil, especially in regions lacking proper containment standards.

– *Mining activities*: mining operations can release heavy metals and other toxic and harmful substances into the soil, leading to contamination of large land areas.

– *Accidental spills*: accidental spills of oil, chemicals or other hazardous materials can result in soil contamination.

4.4.3. *Effects of soil pollution*

Soil pollution can have wide-ranging and significant effects on the environment, ecosystems, human health and agricultural productivity. The following are some of the key effects of soil pollution, which include the following:

1) *Impact on agriculture*: soil pollution can reduce soil fertility, leading to decreased crop yields and poor quality of agricultural produce impacting food security.

2) *Water contamination*: contaminants from polluted soil can leach into groundwater, leading to water pollution and potential health hazards.

3) *Harmful effects on ecosystems*: soil pollution can disrupt the balance of ecosystems, affecting plants, animals and microorganisms.

4) *Environmental contamination*: contaminated soil can lead to the leaching of pollutants into groundwater, posing risks to aquatic ecosystems, drinking water sources and overall environmental integrity. It can also affect the health of plants, animals and microorganisms in the soil.

5) *Health risks*: consumption of crops grown in polluted soil or exposure to contaminated soil can pose health risks to humans and animals through direct contact.

6) *Land degradation*: soil pollution contributes to land degradation, affecting the ability of soil to support vegetation and sustain ecosystem services, leading to desertification, erosion and loss of arable land.

7) *Economic implications*: soil pollution can result in economic losses due to reduced agricultural productivity, increased healthcare costs and the need for remediation efforts.

8) *Biodiversity loss*: soil pollution can disrupt soil ecosystems and reduce biodiversity by harming soil-dwelling organisms, such as earthworms and microorganisms, which play vital roles in nutrient cycling and soil health.

4.4.4. *Mitigation measures of soil pollution*

Soil pollution has a significant environmental impact. The following are some key remedial measures to combat soil pollution, which include the following:

1) *Regulatory measures*: enforce strict regulations on industrial emissions, waste disposal and use of agrochemicals to prevent soil contamination. Implement policies for the proper management and disposal of hazardous waste to minimize soil pollution.

2) *Sustainable agricultural practices*: encourage the use of organic farming methods and integrated pest management to reduce the reliance on chemical pesticides and fertilizers. Promote crop rotation, cover cropping and agroforestry to maintain soil health and fertility.

3) *Soil remediation technique*: implement soil remediation techniques such as phytoremediation (using plants to remove contaminants), bioremediation (using microorganisms to degrade pollutants) and soil washing (removing contaminants from soil through washing) to clean up polluted sites and restore its quality.

4) *Waste management*: establish proper waste management systems and disposal of hazardous waste for industrial, municipal and agricultural waste to prevent the release of pollutants into the soil. Promote recycling and safe disposal of electronic waste, batteries and other hazardous materials to prevent soil pollution.

5) *Monitoring and assessment*: conduct regular soil testing and monitoring to identify polluted areas and assess the extent of contamination. Develop soil quality standards and monitoring programs to track changes in soil health and pollution levels.

6) *Public awareness and education*: raise awareness about the impact of soil pollution on the environment and human health. Educate farmers, industries and the general public about sustainable land use practices and the importance of preserving soil quality.

4.5. Noise pollution

4.5.1. *Meaning and concept*

Noise can be described as any unwanted or undesired sound that disrupts the normal acoustic environment, while noise pollution refers to the presence of excessive or disruptive noise in the environment. Noise pollution refers to the presence of excessive or disturbing noise in the environment that disrupts the normal balance of human or animal life. It can have a range of detrimental effects on physical and psychological well-being as well as on the environment.

In order to fully understand noise pollution, we must take into account its causes, effects, regulations and mitigation techniques. It is one of the ubiquitous risk or hazardous environments that has numerous detrimental effects on wildlife, and that may be harmful to people's health and well-being. There are several negative physical and psychological effects that can result from excessive noise, whether it comes from industrial machines, traffic, building sites or other sources. Extended exposure to loud noises can cause cardiovascular problems, stress, anxiety, sleep disorders and hearing impairment. Furthermore, noise pollution can impair productivity, focus and communication, which lowers the overall quality of life. Regulatory actions, urban planning techniques, engineering controls and personal protection equipment are frequently used to lessen the negative consequences of noise pollution. These initiatives seek to stop noise pollution at its source, stop it from spreading and shield people from prolonged exposure. Public awareness and education campaigns regarding the health risks associated with noise pollution are also crucial in fostering a more noise-conscious society. Through acknowledgement and mitigation of the effects of noise pollution, communities can work toward establishing more tranquil and healthy living spaces for everybody. These efforts aim to reduce noise at the source, block its transmission and protect individuals from excessive exposure. In addition, stakeholders can work toward implementing effective measures to mitigate its impacts, protect human health and well-being and preserve the acoustic environment for current and future generations.

Figure 4.5. *Noise pollution: the invisible danger to our well-being*

4.5.2. *Causes of noise pollution*

1) *Urbanization and industrialization*: rapid urban development and modern civilization lead to increased vehicular traffic, construction activities and industrial operations like machinery, and equipment in factories with high noise intensity, generating noise pollution, and affecting humans and nearby communities.

2) *Transportation*: noise from road traffic, aircraft and railway systems can significantly contribute to noise pollution, especially in densely populated areas.

3) *Recreational activities*: recreational sources such as concerts, events, concerts and festivals with loud music, and also sporting events and leisure activities, can contribute to localized noise pollution.

4) *Construction and demolition*: construction sites and demolition activities produce high levels of noise, affecting surrounding areas.

4.5.3. *Harmful effects of noise pollution*

1) *Health impacts*: prolonged exposure to high levels of noise can lead to increased stress levels, hearing impairment, sleep disturbances, elevated blood pressure, cardiovascular issues, insomnia and other adverse health effects.

2) *Interference with communication*: excessive noise can hinder communication, leading to difficulties in speech perception and comprehension.

3) *Disturbance environment and wildlife*: noise pollution can disrupt the natural behaviors of wildlife, affecting their feeding, mating, and survival patterns, migration patterns and disruption of ecosystems.

4) *Psychological effects*: constant noise exposure can result in irritability, anxiety, negative impact on mental well-being and decreased cognitive performance.

5) *Social impacts*: noise pollution can lead to social conflicts, reduced quality of life in affected areas and community dissatisfaction.

4.5.4. Control measures of noise pollution

1) *Regulatory measures*: the measures include enacting and enforcing noise control regulations and standards for industrial, commercial and residential areas to limit permissible noise levels, and establishing zoning regulations to separate industrial and noisy areas from residential and quiet zones.

2) *Technological solutions*: the solutions include implementing noise reduction technologies for vehicles, machinery and industrial equipment to minimize noise emissions, and encouraging the use of sound barriers, noise-absorbing materials and acoustic design in urban planning and infrastructure development.

3) *Urban planning*: design urban spaces with sound-absorbing landscaping, green buffers, green spaces, sound barriers and strategic placement of buildings to mitigate the impact of noise. Plan transportation routes, residential areas and infrastructure away from noise sources.

4) *Noise control measures*: the measures include implementing sound insulation and double-glazed windows in buildings located in noisy areas to reduce indoor noise levels, and encouraging the use of noise-reducing pavements and traffic management strategies to minimize road traffic noise.

5) *Noise monitoring*: this includes regularly monitoring noise levels in urban areas and taking corrective actions based on monitoring data. In addition, it also includes implementing and enforcing noise control regulations.

6) *Public awareness and education*: this includes raising public/mass awareness about the health impacts of noise pollution and promoting responsible behavior to reduce noise levels. It also includes educating individuals and communities about the importance of maintaining quiet zones and respecting noise regulations, and raising public awareness about the effects of noise pollution.

7) *Community engagement*: this includes encouraging community involvement in identifying noise sources and developing local noise abatement strategies, and establishing community noise action plans and engage stakeholders in noise reduction initiatives.

4.6. Radioactive pollution

4.6.1. *Meaning and nature*

Radioactive pollution refers to the release of any radioactive material into the environment. Radioactive pollution can be extremely dangerous, resulting in abnormal growth and development by mutating DNA and has the potential to cause cancer. Radioactive radiations persist in the environment for years, slowly diminishing over time. There are numerous causes and sources of radioactive pollution, which can have a significant negative impact on human health and the ecosystem. Although there are various methods for controlling this pollution, we still have a long way to go until our planet is free of radioactive waste. Radiation exposure at high levels can cause genetic alterations, cancer and other serious illnesses, among other health problems. Contaminated places may also have negative consequences for agriculture, water sources, ecosystems and biodiversity, all of which affect the balance of the environment as a whole.

4.6.2. *Causes of radioactive pollution*

The primary causes of radioactive pollution include the following:

1) *Nuclear power plants*: accidents in nuclear power plants can lead to the release of radioactive materials into the environment and can cause dangerously high radioactive pollution, such as in the case of Chernobyl nuclear disaster and the Fukushima Daiichi nuclear disaster, which are examples of nuclear power plant accidents.

2) *Improper disposal of radioactive waste*: improper handling and disposal of radioactive waste from industrial, medical and research facilities can result in the contamination of soil and groundwater. This can occur through leaks, spills or inadequate containment measures.

3) *Mining and processing of radioactive materials*: activities related to the mining and processing of radioactive materials such as uranium, phosphate and thorium can generate radioactive waste into the environment. Water runoff, dust discharge and poor waste management can all lead to contamination.

4) *Testing of nuclear weapons*: historic nuclear weapons testing has contributed to radioactive pollution, with fallout from these tests leading to the dispersal of radioactive particles over large areas.

5) *Accidental releases*: accidents during the transportation of radioactive materials, as well as mishandling during industrial processes, can result in the unintended release of radioactive substances into the environment.

6) *Natural sources*: while natural sources of radiation are not considered pollution, certain geological formations can release radioactive substances into the environment. These sources include radon gas from the Earth's crust and radioactive isotopes in mineral deposits.

4.6.3. *Impact of radioactive pollution*

Radioactive pollution can have significant and long-lasting effects on the environment, human health and other living organisms. In conclusion, radioactive pollution can have far-reaching and devastating effects on the environment, human health and socioeconomic well-being. It underscores the importance of stringent regulations, proper waste management and international cooperation to mitigate the risks associated with radioactive pollution. The consequences of radioactive pollution can be severe and can persist for many years. The following are some of the effects of radioactive hazards:

– *Environmental impact*: radioactive substances can contaminate soil and water, affecting the growth of vegetation and posing a risk to both terrestrial and aquatic ecosystems. It can cause genetic mutations in plants, animals and other organisms, leading to abnormalities and reduced reproductive success. In addition, radiation can disrupt ecosystems by affecting the reproductive and developmental processes of plants and animals, leading to imbalances in the food chain and ecosystem stability.

– *Human health impact*: exposure to radioactive pollutants can increase the risk of cancer, particularly leukaemia, thyroid cancer and other solid tumours; it can also lead to genetic disorders in offspring if individuals are exposed to high levels of radiation. Chronic exposure to radioactive pollution can lead to long-term health issues such as cardiovascular diseases, cataracts, and other non-cancer health effects.

– *Social and economic impact*: radioactive pollution can lead to the displacement of communities and the loss of livelihoods, especially in areas directly affected by nuclear accidents or improper disposal of radioactive waste. Treating the health effects of radioactive pollution can result in significant healthcare costs for individuals and governments, impacting economies and the healthcare system.

4.6.4. *Radioactive waste management*

To safeguard both humans and the environment, radioactive waste management refers to the proper handling, storage and disposal of liquid, solid and gas discharge from nuclear industry operations. Any operation that uses nuclear materials, whether for industrial or medicinal purposes, produces waste that is radioactive in one form or another. However, due to its extended lifespan and higher volume of waste production, nuclear energy is the main generator of these wastes. Radioactive wastes need to be handled safely and inexpensively, regardless of where they come from. Generally speaking, radioactive waste is divided into three categories based on its level of radioactivity and the amount of time it remains hazardous: low-level waste (LLW), intermediate-level waste (ILW) and high-level waste (HLW). While most HLW is securely held in designated facilities, disposal of LLW and the majority of ILW is a well-established process. While the scientific and technological community agrees that the permanent disposal of HLW in deep geological repositories is a feasible solution, public society in many nations is still not on board. Various steps can be used to classify the actions required for the management of radioactive waste such as reducing the quantity produced, preparing and wrapping for secure handling and transportation protection, final disposal and temporary storage.

Strategy to lessen radioactive pollution involves strict restrictions for the handling and disposal of radioactive materials. Innovative containment and cleanup solutions are being developed as part of additional efforts to prevent and manage contamination incidents. Understanding radioactive waste management involves comprehensively grasping the processes, challenges and strategies associated with the safe handling, storage and disposal of radioactive materials. The safe and responsible treatment, storage and disposal of materials containing radioactive elements constitute radioactive waste management. It is essential to manage radioactive waste properly to safeguard both the environment and public health. Usually, this procedure entails classifying the waste according to its radioactivity level, safely packing it and then storing or disposing of it in facilities built to stop any radioactive leaks. Creating long-term solutions for managing radioactive waste is a worldwide task that calls for meticulous preparation, close coordination between nations and rigorous respect for safety regulations.

Understanding various aspects of radioactive waste management, stakeholders can work toward developing and implementing effective, science-based strategies to ensure the safe and secure management of radioactive materials, protecting human health and the environment for the long term.

4.7. Plastic pollution

Plastic pollution is a complex and pervasive environmental issue that poses a significant threat to ecosystems, wildlife and human health. The widespread use of plastic products, coupled with inadequate waste management systems, has led to the accumulation of plastic debris in terrestrial and aquatic environments. One of the key challenges of plastic pollution is the longevity of plastic materials in the environment. Most plastics do not biodegrade; instead, they break down into smaller fragments known as microplastics, which can persist in the environment for hundreds to thousands of years. Plastic pollution originates from various sources, including single-use plastics (such as bags, bottles and packaging), microplastics from the breakdown of larger plastic items and industrial discharges of plastic pellets used in manufacturing processes. This persistent pollution not only degrades natural landscapes and harms marine life but also introduces harmful chemicals into the food chain. Addressing plastic pollution requires concerted efforts to reduce single-use plastic consumption, improve recycling infrastructure and promote the development of sustainable alternatives. By raising awareness, implementing regulations and fostering innovation, it is possible to mitigate the impact of plastic pollution and work toward a cleaner, healthier planet for current and future generations.

4.7.1. *Causes of plastic pollution*

1) *Excessive use of single-use plastics*: widespread reliance on single-use plastic items such as bags, bottles and packaging contributes significantly to plastic pollution due to their short lifespan and improper disposal.

2) *Inadequate waste management*: insufficient waste management infrastructure and practices lead to improper disposal of plastic waste,

resulting in littering, landfill overflow and the release of plastics into natural environments.

3) *Lack of recycling and recovery*: limited recycling facilities and low recycling rates lead to a substantial portion of plastic waste ending up in landfills, incinerators or litter, exacerbating the pollution problem.

4) *Plastic microbeads and microfibers*: the use of plastic microbeads in personal care products and the shedding of microfibers from synthetic textiles contribute to microplastic pollution in water bodies.

5) *Industrial and commercial activities*: industrial processes and commercial activities generate plastic waste through packaging, production scrap and discarded products, adding to the overall plastic pollution burden.

6) *Improper disposal and littering*: irresponsible disposal of plastic waste, including littering in urban and natural environments, results in the accumulation of plastic debris in waterways, oceans and on land.

4.7.2. Impact of plastic pollution

Plastic pollution has far-reaching and profound impacts on ecosystems, wildlife, human health and the global environment, from the proliferation of plastic debris in natural habitats to the ingestion of microplastics by marine organisms.

– *Environmental impact*: plastic pollution poses a significant threat to terrestrial and aquatic environments, where plastic debris can persist for hundreds of years, degrading natural landscapes and disrupting ecosystems. The accumulation of plastic waste in oceans, rivers and other water bodies not only harms marine life through entanglement and ingestion but also alters habitats and poses risks to biodiversity.

– *Health impact*: plastic pollution can also have implications for human health. Microplastics, small plastic particles often found in water and soil, have been detected in various food sources, raising concerns about the potential human health effects of plastic ingestion. Additionally, the leaching of harmful chemicals from certain plastics can contaminate soil and water, with potential implications for agricultural productivity and human health. Chemicals found in plastics, such as phthalates and bisphenol A (BPA), can leach into the environment and accumulate in the food chain, potentially posing health risks to humans through the consumption of contaminated seafood or water.

– *Economic costs*: plastic pollution imposes economic costs on society, including costs associated with clean-up efforts, damage to tourism and fisheries industries and impacts on marine transportation and infrastructure.

– *Global distribution*: plastic pollution is a global issue affecting oceans, rivers, lakes and even remote and pristine environments. It is estimated that millions of tons of plastic enter the oceans annually, carried by ocean currents and wind.

4.7.3. *Mitigation strategies*

Mitigating plastic pollution requires a multifaceted approach. Strategies include reducing plastic consumption through waste reduction initiatives, promoting recycling and circular economy models, improving waste management infrastructure, implementing policies to ban or regulate single-use plastics and raising awareness about the environmental impacts of plastic pollution. Plastic pollution is a significant environmental concern that requires effective mitigation measures to address its impact on ecosystems and human health. Several strategies and interventions have been proposed and implemented to mitigate plastic pollution. The following are some key mitigation measures:

1) *Reduce, reuse and recycle (3Rs)*: this encourages the adoption of the 3Rs approach to minimize the generation of plastic waste, promotes the re-use of plastic products and materials to extend their lifespan, and supports recycling initiatives to divert plastic waste from landfills and reduce the demand for new plastic production.

2) *Single-use plastic bans and regulations*: this implements bans or restrictions on single-use plastics such as plastic bags, straws and utensils to reduce their consumption and subsequent pollution.

3) *Promotion of alternative materials*: this encourages the use of biodegradable and compostable alternatives to traditional plastics, and supports the development and adoption of sustainable packaging materials to replace single-use plastics.

4) *Waste management and infrastructure*: this invests in efficient waste management infrastructure to improve the collection, sorting and recycling

of plastic waste, and implements proper disposal methods and facilities to prevent plastic waste from entering water bodies and natural habitats.

5) *Public awareness and education*: this conducts educational campaigns to raise awareness about the environmental impact of plastic pollution, and promotes responsible consumer behavior and advocating for sustainable lifestyle choices.

6) *Innovation and technology*: this invests in research and development of innovative technologies for plastic recycling and upcycling, and supports initiatives that focus on developing plastic alternatives and improving the recyclability of plastics.

7) *Clean-up and restoration efforts*: this organizes and participates in community-based clean-up activities to remove plastic debris from beaches, rivers and other natural environments, and supports restoration projects to rehabilitate ecosystems affected by plastic pollution.

4.8. Questions

1) What are the consequences of air pollution on life and the environment?

2) Describe the causes and consequences of air pollution in urban areas.

3) What are the different types of pollutants and their sources? What are their impacts on human health and the environment?

4) Describe the sources of air pollution in urban areas, and how it affects human health.

5) What are the long-term consequences of air pollution on climate and ecosystems?

6) Define pollution and industrialization.

7) How do industrial activities contribute to air pollution, and what are the most significant pollutants emitted from these sources?

8) How does air pollution impact developing countries differently from developed nations, and what are the unique challenges they face in addressing this issue?

9) What are the short-term and long-term health effects of air pollution on individuals, particularly those living in highly polluted cities?

10) How can individuals contribute to mitigating air pollution in their daily lives?

11) Define water pollution. What are the major anthropogenic causes responsible for it?

12) What are the main sources of water pollution, and how do human activities contribute to the contamination of freshwater bodies such as rivers, lakes and groundwater?

13) What role do agricultural practices, industrial discharges and urban runoff play in causing water pollution, and what specific pollutants are commonly associated with these sources?

14) What are the potential ripple effects of water pollution on food chains, and how does it disrupt the balance of aquatic habitats?

15) How can communities and individuals contribute to preventing water pollution in their local areas?

16) How does prolonged exposure to high levels of noise pollution impact human health, including both physical and psychological effects?

17) What is noise pollution and its main causes?

18) What measures can be taken to mitigate noise pollution from transportation sources such as aircraft, road traffic and railways?

19) How does noise pollution impact the quality of life in residential areas, and what are the social and economic implications of living in noisy environments?

20) What are some strategies for creating more acoustically friendly urban spaces and buildings to minimize the negative effects of noise pollution on residents?

21) How can individuals and communities actively participate in noise pollution reduction efforts?

22) What are some of the best practices for raising awareness about the importance of sound management in urban environments?

23) What is radioactive pollution? What are the primary sources of radioactive pollution?

24) How do human activities contribute to the release of radioactive substances into the environment?

25) What are the health risks associated with exposure to radioactive pollution, and how do these risks vary depending on the type of radioactive substance and the duration of exposure?

26) What are the primary sources of radioactive pollution, and how do human activities contribute to the release of radioactive substances into the environment?

27) What is soil pollution? What are the primary sources of radioactive pollution?

28) How does soil pollution affect agricultural productivity and the quality of food crops, and what are the potential risks associated with consuming produce grown in contaminated soils?

29) What are the best practices for preventing soil pollution, including sustainable waste management and pollution control measures?

30) What are the long-term consequences of soil pollution on biodiversity?

4.9. Multiple-choice questions (MCQs)

1) What is the primary source of air pollution in urban areas?

a) Industrial emissions

b) **Vehicle exhaust**

c) Agricultural activities

d) Natural sources

2) Which gas is a major component of smog and contributes to respiratory problems?

a) Oxygen

b) Carbon monoxide

c) **Nitrogen dioxide**

d) Methane

3) What is the main cause of water pollution in rivers and lakes?

a) **Agricultural runoff**

b) Industrial discharges

c) Oil spills

d) Natural sedimentation

4) Which of the following is a greenhouse gas responsible for global warming?

a) Oxygen

b) **Carbon dioxide**

c) Nitrous oxide

d) Hydrogen

5) What is the primary cause of noise pollution in urban areas?

a) Construction activities

b) **Traffic noise**

c) Industrial machinery

d) Loud music

6) Which pollutant is a result of incomplete combustion of fossil fuels?

a) Sulfur dioxide

b) Particulate matter

c) **Carbon monoxide**

d) Lead

7) What is the main source of plastic pollution in the oceans?

a) Industrial waste

b) Fishing nets

c) **Single-use plastics**

d) Natural debris

8) What is the primary cause of soil pollution in agricultural areas?

a) **Pesticides and herbicides**

b) Industrial waste

c) Deforestation

d) Mining activities

9) Which of the following is a key factor contributing to indoor air pollution?

a) **Tobacco smoke**

b) Vehicle emissions

c) Industrial waste

d) Agricultural runoff

10) What is eutrophication, a phenomenon associated with water pollution?

a) Accumulation of heavy metals

b) **Excessive nutrient enrichment**

c) Oil contamination

d) Radioactive waste

11) Which air pollutant is responsible for the formation of acid rain?

a) Carbon monoxide

b) **Sulfur dioxide**

c) Nitrogen oxide

d) Ozone

12) What is the primary source of mercury pollution in aquatic ecosystems?

a) **Coal-fired power plants**

b) Nuclear power plants

c) Natural volcanic activity

d) Agricultural runoff

13) What is the main cause of thermal pollution in water bodies?

a) **Industrial discharges**

b) Urban runoff

c) Oil spills

d) Nuclear accidents

14) Which pollutant is associated with the depletion of the ozone layer?

a) **Chlorofluorocarbons (CFCs)**

b) Carbon dioxide

c) Methane

d) Nitrous oxide

15) What is the primary source of noise pollution in natural habitats?

a) Wildlife sounds

b) **Human activities**

c) Wind

d) Earthquakes

16) What is the primary cause of radiation pollution?

a) **Nuclear accidents**

b) Solar radiation

c) X-rays

d) Volcanic activity

17) Which of the following is a major contributor to light pollution?

a) **Streetlights**

b) Natural moonlight

c) Stars

d) Vehicle headlights

18) What is the primary source of nutrient pollution in water bodies?

a) Industrial discharges

b) **Agricultural runoff**

c) Oil spills

d) Urban wastewater

19) Which greenhouse gas is released during the decay of organic matter in landfills?

a) Carbon dioxide

b) **Methane**

c) Nitrous oxide

d) Water vapor

20) What is the primary source of lead pollution in urban environments?

a) Vehicle emissions

b) Industrial processes

c) **Paint**

d) Nuclear power plants

21) Which of the following is a common indoor air pollutant?

a) **Radon**

b) Carbon dioxide

c) Ozone

d) Sulfur dioxide

22) What is the primary cause of deforestation that contributes to air and soil pollution?

a) **Agricultural expansion**

b) Urbanization

c) Logging

d) Wildfires

23) Which pollutant is associated with the phenomenon of "acidification" in aquatic ecosystems?

a) Carbon monoxide

b) **Sulfur dioxide**

c) Nitrogen oxide

d) Lead

24) What is the primary source of thermal pollution in rivers and lakes?

a) **Industrial discharges**

b) Urban runoff

c) Agricultural runoff

d) Nuclear power plants

25) Which gas contributes to the formation of ground-level ozone, a major air pollutant?

a) Carbon dioxide

b) **Nitrogen dioxide**

c) Methane

d) Hydrogen

26) What is the primary cause of marine pollution by oil spills?

a) Industrial discharges

b) **Shipping activities**

c) Agricultural runoff

d) Natural seepage

27) Which pollutant is a byproduct of electronic waste (e-waste) burning?

a) Lead

b) Mercury

c) Cadmium

d) **All of the above**

28) What is the primary source of nutrient pollution in coastal areas?

a) Agricultural runoff

b) **Urban wastewater**

c) Industrial discharges

d) Marine activities

29) Which of the following pollutants contributes to the formation of photochemical smog?

a) Sulfur dioxide

b) **Nitrogen oxides**

c) Carbon monoxide

d) Particulate matter

30) The Ebola virus disease first appeared in the year:

a) **1976**

b) 1967

c) 1956

d) 1926

5

Environmental Challenges

Environmental challenges encompass the obstacles and difficulties stemming from human or anthropogenic activities and natural processes, which have adverse effects on ecosystems, biodiversity and the overall environment. These challenges pose significant threats to the well-being of our planet, ecosystems and communities. Some of the key environmental challenges that need to be addressed at local, national and global levels include climate change, poor waste management, water scarcity, waterlogging, sea level rise, water pollution, desertification, biodiversity loss, land degradation, global warming, greenhouse effect, ocean, acidification, ozone depletion and genetically modified foods. These challenges make us vulnerable to disasters both now and in the future. Throughout history, humans have faced environmental challenges, but the industrialization and urbanization processes have exacerbated these issues, for example, air and water pollution from factories and urban living conditions cause significant problems. Concern over pollution quickly expanded to encompass a range of environmental conditions. As globalization continues to blur borders, local environmental problems are transforming into international issues. The world is currently facing major environmental problems, such as acid rain, air pollution, hazardous wastes, ozone layer depletion, greenhouse effect, water pollution, overpopulation and deforestation. These issues not only harm the environment but also have detrimental effects on all living beings on Earth.

One of the main reasons for the increase in environmental disasters is human activity. Deforestation, pollution and our continuous use of natural resources are all seriously disrupting the delicate balance of our ecosystem. Climate change, driven by greenhouse gas (GHG) emissions, is also exacerbating the situation, leading to more frequent and severe disasters such as droughts, floods and wildfires. Another critical factor is the rapid urbanization and industrialization taking place around the world. As more and more people move to cities and industries expand, the demand for land and resources is skyrocketing. This often leads to the destruction of habitats, loss of biodiversity and increased vulnerability to natural disasters.

Furthermore, inadequate environmental policies and regulations play a significant role in the escalation of environmental challenges. The lack of strict enforcement and accountability allows companies to continue harmful practices, putting our ecosystem at risk.

Despite these challenges, there are ways that people can work together to address and overcome environmental issues and create solutions that will benefit both the environment and society as a whole. By raising awareness, advocating for policy changes and adopting sustainable practices in our daily lives, individuals can make a positive impact on the environment and help create a healthier future for all. We must take action now to address these issues before it is too late. We must prioritize sustainable practices, protect and restore our natural habitats and reduce our carbon footprint. Investing in renewable energy sources, promoting conservation efforts and implementing strict regulations are crucial steps in safeguarding our ecosystem from further disasters and challenges.

Environmental challenges faced by people

Environmental challenges have become a pressing issue in today's society. People all around the world are experiencing numerous problems as a result of these challenges. From pollution to climate change, the impacts of environmental issues are being felt by individuals in various ways.

The first challenge that many people face is pollution. The contamination of air, water and soil is causing numerous health problems for individuals. People living in urban areas often have to deal with poor air quality, leading to respiratory issues and other health concerns. Additionally, the pollution of water sources is causing waterborne diseases and other health problems for communities around the world.

Climate change is another significant environmental challenge. The rising global temperatures are leading to more frequent and severe natural disasters, such as hurricanes, floods and wildfires. These disasters are causing widespread destruction and displacement of communities, leaving people vulnerable and in need of support.

In addition to pollution and climate change, deforestation and loss of biodiversity are also major challenges faced by people. The destruction of forests and habitats is leading to the loss of plant and animal species at an alarming rate. This loss of biodiversity can have far-reaching consequences for ecosystems and human populations alike.

To combat these environmental challenges and create a sustainable future, it is essential to implement sustainable practices, invest in renewable energy, conserve natural habitats, promote environmental education and adopt policies that prioritize environmental protection and conservation. Coordinated efforts from governments, businesses, communities and individuals are crucial in overcoming these challenges and preserving our planet for future generations.

5.1. Climate change

Climate change is one of the most significant challenges facing humanity in the 21st century. It is a global phenomenon characterized by the alteration of Earth's climate patterns, primarily attributed to human activities such as the burning of fossil fuels deforestation, and industrial processes. Climate change leads to rising global temperatures, extreme weather events, sea-level rise, waterlogging in urban and rural areas and disruptions to ecosystems and human societies. The atmospheric release of GHGs is the main cause of climate change. Methane (CH_4), nitrous oxide (N_2O) and carbon dioxide (CO_2) are a few of the most important GHGs released into the atmosphere by human activity. The main source of CO_2 emissions is the combustion of fossil fuels for industrial activities, transportation, and energy production. Methane and nitrous oxide emissions are the main GHG emissions from cattle farming, agriculture and deforestation.

5.1.1. *Causes of climate change*

1) *GHG emissions*: the burning of fossil fuels (coal, oil and natural gas) for energy production, transportation, industry and agriculture releases large amounts of GHGs into the atmosphere. Among these gases, nitrous oxide (N_2O), carbon dioxide (CO_2), and methane (CH_4) are the most important. These gases trap heat in the Earth's atmosphere, leading to global warming and climate change.

2) *Deforestation and land use changes*: deforestation, primarily for agriculture, and urban development reduces the Earth's capacity to absorb CO_2 from the atmosphere through photosynthesis. Additionally, land use changes, such as converting forests or grasslands into agricultural land, can release stored carbon into the atmosphere.

3) *Industrial processes*: certain industrial processes release potent GHGs, such as hydrofluorocarbons (HFCs), perfluorocarbons (PFCs) and sulfur hexafluoride (SF_6). These gases are used in various industrial applications, including refrigeration, air conditioning, manufacturing and semiconductor production.

4) *Agricultural practices*: agricultural activities, including livestock farming, rice cultivation, and the use of synthetic fertilizers, produce methane and nitrous oxide emissions. Livestock, particularly cattle, produce methane during digestion (enteric fermentation), while nitrogen fertilizers contribute to nitrous oxide emissions through microbial processes in soil.

5) *Waste management*: improper waste management practices, such as open burning of waste, landfill decomposition and wastewater treatment, can release methane and CO_2 into the atmosphere.

6) *Urbanization*: rapid urbanization and associated activities, including increased energy consumption, transportation emissions and changes in land use patterns, contribute to localized climate effects, such as the urban heat island effect.

5.1.2. *Impact of climate change on global sustainability*

Climate change poses significant and wide-ranging impacts on ecosystems, economies and societies worldwide. Rising temperatures lead to melting ice caps and glaciers, contributing to sea-level rise and threatening coastal communities with increased flooding and erosion. Shifts in precipitation patterns result in more frequent and intense droughts, heatwaves and wildfires in some regions, while others experience heavier rainfall and flooding. These changes disrupt agriculture, water resources and biodiversity, endangering food security and ecosystem stability.

Climate change leads to an overall increase in global temperatures, resulting in heatwaves, altered weather patterns and changes in the distribution of plant and animal species. Additionally, climate-related extreme weather events exacerbate socioeconomic inequalities, disproportionately affecting vulnerable populations and exacerbating poverty and displacement. Furthermore, climate change exacerbates health risks, including heat-related illnesses, vector-borne diseases and respiratory problems due to air pollution.

The consequences of climate change are wide-ranging and profound, affecting ecosystems, economies and human well-being. Rising global temperatures lead to more frequent and intense heatwaves, droughts and wildfires, threatening agricultural productivity and food security. Sea level rise is a result of melting ice caps and glaciers, putting coastal ecosystems and communities in danger. Changes in precipitation patterns disrupt water resources, exacerbating water scarcity in some regions and increasing the risk of floods in others. Climate change also contributes to the spread of vector-borne diseases, heat-related illnesses and respiratory problems, affecting human health and placing additional burdens on healthcare systems. Biodiversity loss, ocean acidification and extreme weather events further enhance the impacts of climate change, posing significant challenges to sustainable development. By understanding these effects of climate change, policymakers, businesses and communities can work toward implementing adaptation and mitigation measures to address the impacts of climate change and build resilience in the face of these challenges.

5.1.3. *Solution to combat climate change*

Combating climate change requires a comprehensive and coordinated response at the local, national and international levels, which involves reducing GHG emissions, transitioning to renewable energy sources, promoting sustainable practices and enhancing resilience to climate impacts. Transitioning to renewable energy sources such as solar, wind, and hydroelectric power is essential to reduce CO_2 emissions from the energy sector. Improving energy efficiency in buildings, transportation and industries can also significantly contribute to emissions reduction. Sustainable land management practices, reforestation efforts and conservation initiatives are crucial for preserving ecosystems and sequestering carbon. Additionally, implementing policies and regulations to incentivize emission reductions, promote sustainable practices and foster innovation in clean technologies is essential for achieving long-term climate goals.

Some key measures to combat climate change include the following:

1) *Transition to renewable energy*: accelerate the transition from fossil fuels to renewable energy sources such as solar, wind, hydroelectric and geothermal energy. This involves increasing investment in renewable energy infrastructure, improving energy efficiency and phasing out subsidies for fossil fuels.

2) *Energy efficiency*: implement energy efficiency measures across various sectors, including buildings, transportation, industry and agriculture. This includes improving building insulation, upgrading transportation systems and adopting energy-efficient technologies and practices in manufacturing and agriculture.

3) *Regulatory policies*: implement and strengthen regulatory policies to limit GHG emissions, such as carbon pricing mechanisms, emission standards for vehicles and industries, and renewable energy mandates. International agreements like the Paris Agreement play a crucial role in coordinating efforts to combat climate change.

4) *Afforestation and reforestation*: increase afforestation (planting trees in areas that have not been forested before) and reforestation (replanting trees in areas that have been deforested) efforts to enhance carbon sequestration and biodiversity conservation. Protecting existing forests from deforestation and degradation is also essential.

5) *Sustainable land use*: promote sustainable land management practices, including sustainable agriculture, agroforestry and conservation agriculture, to reduce emissions from land-use change and increase carbon sequestration in soils.

6) *Education and awareness*: raise awareness and educate the public about climate change, its impacts and the importance of taking action.

This includes promoting sustainable lifestyles, encouraging climate-friendly consumer choices and fostering a culture of environmental stewardship.

5.2. Greenhouse effect

The greenhouse effect is a phenomenon where certain gases in the Earth's atmosphere, such as carbon dioxide (CO_2), methane (CH_4), nitrous oxide (N_2O) and water vapour, trap heat from the Sun. This prevents the heat from escaping back into space, leading to a warming effect on the planet.

The greenhouse effect plays a crucial role in regulating Earth's climate by trapping heat in the atmosphere, making our planet suitable for life. However, human activities have exacerbated this natural process, resulting in global warming and climate change. GHGs are the primary drivers of the greenhouse effect. When solar radiation enters the Earth's atmosphere, it heats the surface. Some of this heat is absorbed by the surface and re-emitted as infrared radiation. GHGs then absorb and re-radiate this infrared radiation, trapping the heat in the atmosphere and warming the lower atmosphere and Earth's surface. The main GHGs responsible for the greenhouse effect are carbon dioxide (CO_2), methane (CH_4), nitrous oxide (N_2O), and water vapour. These gases play a crucial role in regulating Earth's climate and maintaining average global temperatures conducive to life. Human activities, such as burning fossil fuels, deforestation, industrial processes, and agriculture, have significantly increased the concentration of GHGs in the atmosphere. This enhanced greenhouse effect has resulted in global warming and climate change, impacting ecosystems and weather patterns. Understanding the greenhouse effect and its impact on global climate is essential in addressing the challenges of climate change. By reducing GHG emissions and promoting sustainable practices, we can help mitigate the effects of global warming and ensure a habitable planet for future generations. GHGs absorb infrared radiation emitted by the Earth's surface, preventing it from escaping into space and leading to the warming of the lower atmosphere and Earth's surface. While the greenhouse effect is a natural phenomenon necessary for maintaining Earth's temperature within a habitable range, human activities have significantly intensified it by increasing the concentration of GHGs in the environment, leading to global warming and climate change. The greenhouse effect, a natural phenomenon crucial for maintaining Earth's temperature within a habitable range, has been significantly amplified by human activities, leading to profound impacts on the planet's climate system.

5.2.1. *Factors contributing to greenhouse effect*

The Earth's surface warms naturally due to the greenhouse effect, but humans have this phenomenon, leading to overall climate change. The following are the causes of the greenhouse effect:

1) *GHGs*: the emission of GHGs like carbon dioxide (CO_2), methane (CH_4), nitrous oxide (N_2O), and fluorinated gases from human activities such as burning fossil fuels, deforestation, industrial processes and agriculture increases the concentration of these gases in the atmosphere, trapping more heat.

2) *Deforestation*: the cutting down of forests reduces the number of trees available to absorb CO_2 from the atmosphere, leading to an increase in GHGs.

3) *Industrialization*: industrial processes produce large amounts of GHGs, such as CO_2 from burning fossil fuels for energy and methane from activities like mining and agriculture. Various industrial and residential activities release fluorinated gases, which have a high global warming potential.

4) *Urbanization*: urban development leads to the creation of heat islands, where concrete and asphalt absorb and re-radiate the heat, contributing to local and regional warming.

5) *Agricultural practices*: farming activities, including livestock rearing and rice paddies, release methane, a potent GHG.

6) *Use of fossil fuels*: burning coal, oil and natural gas for energy and transportation releases CO_2 into the atmosphere, contributing to the greenhouse effect.

7) *Waste management*: improper disposal of waste, particularly in landfills, leads to the release of methane as organic waste decomposes.

5.2.2. *Impact of greenhouse effect*

One of the most evident effects of the greenhouse effect is global warming, characterized by a steady increase in the Earth's average surface temperature. This warming trend is primarily attributed to the enhanced greenhouse effect caused by human activities, particularly the release of GHGs such as carbon dioxide (CO_2), methane (CH_4) and nitrous oxide (N_2O) from burning fossil fuels, deforestation, agriculture and industrial processes. As these gases accumulate in the atmosphere, they trap heat radiated from the Earth's surface, resulting in a rise in global temperatures. Climate change, driven by the greenhouse effect, also exacerbates extreme weather events such as heatwaves, droughts, wildfires, hurricanes and heavy rainfall events. These events have devastating consequences for human populations, including loss of life, displacement of communities, destruction of infrastructure and disruptions to agriculture, water resources and economies. Vulnerable populations, including low-income communities, indigenous peoples and marginalized groups, are disproportionately affected by these climate-related disasters, exacerbating social inequalities and increasing the risk of conflict and displacement.

Furthermore, the greenhouse effect contributes to various feedback mechanisms that amplify climate change. For example, as polar ice melts, it reduces the Earth's albedo, or reflectivity, leading to more absorption of solar radiation and further warming. Similarly, the thawing of permafrost releases stored carbon and methane, intensifying GHG emissions and accelerating climate change. These feedback loops underscore the urgency of addressing the root causes of the greenhouse effect to mitigate its impacts and prevent runaway climate change.

In addition to environmental and social consequences, the greenhouse effect poses significant economic risks and challenges. Climate-related disasters and extreme weather events impose substantial costs on economies, including damages to infrastructure, disruptions to supply chains, losses in agricultural productivity, increased healthcare expenditures and impacts on tourism and recreation industries. Moreover, the long-term consequences of climate change, such as sea-level rise, biodiversity loss, and ecosystem degradation, pose significant economic risks, including declining property values, insurance liabilities and impacts on investments and financial markets.

Increased GHG emissions have led to a rise in average global temperatures, resulting in changes to weather patterns, melting ice caps and rising sea levels. The intensified greenhouse effect contributes to more frequent and severe heatwaves, storms, floods and droughts. Climate change disrupts ecosystems, leading to shifts in species distributions, loss of biodiversity and degradation of natural habitats.

5.2.3. *Effective strategies for reducing GHG emissions*

Global climate change is a result of the greenhouse effect, which is brought on by the build-up of GHGs in the Earth's atmosphere. It is imperative to put in place several policies targeted at cutting GHG emissions and encouraging sustainable behaviors to minimize this problem. It includes various actions as discussed below:

– *Renewable energy sources*: utilizing solar, wind and hydropower has the potential to greatly diminish our dependence on fossil fuels, consequently lowering emissions and combating climate change.

– *Enhancing energy efficiency*: implementing energy-efficient technologies across various sectors, including industries, transportation and households, can result in significant energy savings and a subsequent reduction in emissions.

– *Energy efficiency*: implementing energy-efficient technologies in industries, transportation and households can lead to substantial reductions in energy consumption and, consequently, lower GHG emissions.

– *Emphasizing afforestation*: this includes planting trees and restoring forests, as they play a very crucial role in absorbing carbon dioxide, acting as a natural carbon sink to minimize the greenhouse effect.

– *Enforcing regulations and policies*: GHG emissions can be effectively controlled by supporting carbon pricing mechanisms and enforcing strict rules and policies aimed at reducing emissions from transportation, agricultural and industrial operations.

– *Promoting sustainable practices*: encouraging and promoting sustainable land use, advocating for organic farming and embracing eco-friendly practices in various sectors can collectively contribute to mitigating the greenhouse effect.

5.3. Ozone depletion

The shrinking of the ozone layer in the Earth's stratosphere is known as "ozone depletion". As the ozone layer absorbs most of the Sun's harmful ultraviolet (UV) radiation, it plays a critical role in preserving life on Earth. However, human-made chemicals called ozone-depleting substances (ODS), such as chlorofluorocarbons (CFCs), halons and other industrial gases, have been released into the atmosphere. These ODS molecules eventually reach the stratosphere, where they are broken down by solar radiation, releasing chlorine and bromine atoms. These chlorine and bromine atoms then catalyze the destruction of ozone molecules, leading to the thinning of the ozone layer. This thinning allows more UV radiation to reach the Earth's surface, posing significant risks to human health, ecosystems and the environment.

The most significant ozone depletion occurs in the Polar Regions, resulting in the formation of the ozone hole, particularly over Antarctica. Ozone depletion also occurs to a lesser extent in other parts of the world.

Efforts to address ozone depletion have been largely successful through international cooperation, most notably with the implementation of the Montreal Protocol in 1987. This treaty aimed to phase out the production and consumption of ODS, leading to a gradual recovery of the ozone layer. However, continued vigilance and adherence to the protocols outlined in the Montreal Protocol are necessary to ensure the long-term recovery of the ozone layer and the protection of the Earth's environment. This layer also consumes a large part of the harmful ultraviolet radiation in the Sun's rays and thus saves human, animal, and plant life on the Earth from the harmful effects of ultraviolet radiation. The entire process of ozone depletion and its effects are more complicated than the simple description of ozone layer depletion. Ozone molecules are destroyed by the atoms of chlorine and bromine in the atmosphere that come into contact with them. When certain substances are exposed to intense UV light, they emit bromine and chlorine, which further depletes the ozone layer. The depleting substances that contain chlorine, including CFCs, are the most abundant ODS.

5.3.1. *Causes of ozone depletion*

There are numerous factors linked to the depletion of the ozone layer, which is a significant problem. The main causes of ozone depletion can be discussed as follows:

– *Chlorofluorocarbons*: the primary cause of the ozone layer's thinning is CFCs. These arc released by air conditioners, freezers, spray aerosols and other machines.

– Chlorine atoms are released when ultraviolet radiation breaks down the CFC molecules in the stratosphere. Ozone is destroyed by the reaction of these atoms in it.

– *Uncontrolled rocket launches*: according to research, the uncontrolled launching of rockets results in more ozone layer depletion than CFCs. If unchecked, there may be considerable depletion of ozone layer by 2050.

– *Nitrogenous molecules*: the ozone layer's thinning is mostly caused by nitrogenous molecules, including NO_2, NO and N_2O.

– *Natural causes*: it has been discovered that several natural processes, such as sunspots and stratospheric haze, cause the ozone layer to thin.

5.3.2. *Impact of ozone layer depletion*

There are negative consequences of ozone layer depletion, especially for the ecosystem. Let us see the major impact of ozone layer depletion on humans and the environment.

– *Impact on human health*: if the ozone layer is depleted, humans will be directly exposed to the harmful ultraviolet rays of the Sun. Direct exposure to ultraviolet radiation might result in serious health issues among humans and could lead to skin diseases like cancer, sunburns, cataracts and quick ageing.

– *Impact on the environment*: intense UV radiation can cause minimal plant growth and photosynthesis. The damaging effects of UV rays are also felt by forest land areas.

– *Impact on marine life*: planktons are greatly affected by exposure to harmful ultraviolet rays. Their concentration is higher in the aquatic food chain. The creatures in the lower food chain are also affected if planktons are destroyed.

– *Climate change*: this is caused by altering atmospheric circulation patterns and affecting the distribution of ozone and other GHGs. Changes in atmospheric circulation can influence regional climate variability, including shifts in precipitation patterns, temperature extremes and weather events. Furthermore, ODS are potent GHGs, contributing to global warming and climate change.

5.3.3. *Mitigation measures*

Ozone layer depletion has a huge impact on humans and the environment. It is thus crucial to reduce its effect at the global level. The following present some ideas to aid in the worldwide prevention of this issue:

1) *Avoid utilizing pesticides*: rather than utilizing chemicals to eradicate weeds and pests, natural approaches should be used. To get rid of pests, environmentally safe insecticides can be used, or weeds can be pulled out by hand. Most cleaning products contain chemicals that damage the ozone layer by releasing chlorine and bromine into the atmosphere. Natural materials should be used instead to protect the environment.

2) *Promotion of sustainable practices*: encourage sustainable practices that reduce the emission of ODS, such as energy efficiency measures and waste reduction.

3) *Substitution of ODS*: encourage the phase-out of ODS such as CFCs, halons and hydrochlorofluorocarbons (HCFCs) by promoting the use of environmentally friendly alternatives such as HFCs with low ozone depletion potential.

4) *The use of nitrous oxide should be prohibited*: it is time for the government to step in and restrict the use of hazardous nitrous oxides, which are having a negative impact on the ozone layer. In order to also reduce its use on individual levels, people should be made aware of the negative consequences of nitrous oxide and the products that emit this gas.

5) *Regulatory measures*: implement and enforce international agreements and regulations aimed at reducing the production and use of ODS. The Montreal Protocol is a prime example of successful international cooperation in this regard.

6) *Public awareness and education*: increase public awareness about the ozone layer depletion and its consequences. Education campaigns can encourage individuals and businesses to adopt ozone-friendly practices and products.

5.4. Global warming

The average surface temperature of the Earth has been rising over a prolonged period as a result of human activity, mainly the release of GHGs such as carbon dioxide (CO_2), methane (CH_4) and nitrous oxide (N_2O). Temperatures gradually rise as a result of these gases' ability to trap heat in the Earth's atmosphere. A rise in GHGs, such as carbon dioxide, methane and nitrous oxide, slows down the Earth's surface-nighttime cooling, which raises the minimum temperature (morning temperature) nearly everywhere on Earth and gives rise to the phenomena known as global warming. The atmosphere warms as a result of heat, which is trapped due to the greenhouse effect. Global warming is the unusually rapid increase in the Earth's average temperature over the past century, primarily due to the GHGs released by people burning fossil fuels. Global warming happens when sunlight and solar radiation that have reflected off the Earth's surface are absorbed by carbon dioxide (CO_2), other air pollutants and GHGs that have accumulated in the atmosphere. The globe then warms up as a result of these contaminants' ability to trap heat in the atmosphere for years or even centuries. Normally, this radiation would escape into space. Global warming has far-reaching and significant effects. Rising sea levels are one of the most obvious effects, which are caused by the melting of polar ice caps and glaciers. This phenomenon threatens coastal communities worldwide, exacerbating the risk of flooding, erosion and saltwater intrusion into freshwater sources. Moreover, changes in temperature and precipitation patterns disrupt ecosystems, causing shifts in species distributions, alterations in habitat suitability and disruptions to ecological processes such as migration, reproduction and food availability. Global warming is a pressing issue that requires effective control measures to mitigate its impact. Several strategies can be employed to address global warming, ranging from policy initiatives to individual actions. Implementation of effective mitigation measures at global, national and local levels is crucial in addressing the challenges of global warming and working toward a more sustainable future.

5.4.1. *Causes of global warming*

The issues that cause global warming include human influences such as:

– *GHGs*: GHGs consist of carbon monoxide and sulfur dioxide. They trap solar heat rays and prevent them from escaping from the surface of the Earth. This has caused Earth's temperature to increase.

– *Volcanic eruptions*: this is another issue that causes global warming. For example, the amount of ash and carbon dioxide released into the atmosphere during a single volcanic eruption.

– *Mining activities*: during the process of mining, methane will be trapped below the Earth. Additionally, rearing cattle releases manure, which is another source of methane.

– *Deforestation*: the most common issue is deforestation because humans have been cutting down trees to produce paper, wood, build houses and more.

5.4.2. *Effects of global warming*

The impacts of global warming are far-reaching and affect various aspects of the environment, ecosystems, and human societies:

– *High temperatures*: higher temperatures cause polar ice caps and glaciers to melt at accelerated rates. This contributes to rising sea levels, which pose risks to coastal communities, habitats and infrastructure. It also disrupts ecosystems dependent on ice, such as polar bear habitats and Arctic ecosystems. Once the ice melts, the first effect will be rise in sea levels because the melting glaciers become oceans. This poses risks to low-lying coastal areas, increasing the likelihood of coastal erosion, flooding and saltwater intrusion into freshwater sources.

– *Loss of habitat and biodiversity*: as a result of climate change, species such as tropical frogs and polar bears are expected to become extinct. Global warming exacerbates habitat loss, fragmentation and degradation, leading to a decline in biodiversity and disruption of the ecosystem.

– *Changes in precipitation patterns*: global warming alters precipitation patterns, leading to changes in rainfall distribution, intensity and frequency.

This can result in more frequent and severe droughts, floods and storms, impacting agriculture, water resources and infrastructure.

– *Ocean acidification*: increased atmospheric CO_2 levels lead to higher levels of CO_2 being absorbed by the world's oceans, resulting in ocean acidification. This has harmful effects on marine ecosystems, particularly coral reefs and shell-forming organisms, which depend on carbonate ions for their skeletons and shells.

– *Impact on agriculture*: changes in temperature and precipitation patterns affect agricultural productivity and food security. Shifts in growing seasons, increased frequency of extreme weather events and changes in pest and disease patterns can lead to reduced crop yields and livestock productivity.

– *Health impacts*: global warming contributes to various health risks, including heat-related illnesses, respiratory problems from air pollution, vector-borne diseases like malaria and dengue fever spreading to new areas, and food and waterborne diseases due to changing conditions.

5.4.3. *Control measures*

Governments and people must act now to put global warming solutions into practice. We can take steps to lessen the amount of GHGs that enter the atmosphere in order to prevent global warming.

– *Reduction of GHG emissions*: encouraging the use of solar, wind, hydro and other renewable energy sources to reduce reliance on fossil fuels. Improving energy efficiency in industries, transportation and buildings to lower overall energy consumption and subsequent emissions.

– *Afforestation*: planting trees in areas that were previously devoid of any forest cover, restoring and replanting trees in areas that have been deforested.

– *Recycling*: recycling can reduce garbage by reusing plastic bags, bottles, papers or glass.

– *Reduction of GHG emissions*: encouraging the use of solar, wind, hydro and other renewable energy sources to reduce reliance on fossil fuels. Improving energy efficiency in industries, transportation and buildings to lower overall energy consumption and subsequent emissions.

– *Sustainable agriculture and land use*: promoting sustainable agricultural practices that reduce emissions and sequester carbon in soil. Protecting and restoring ecosystems like wetlands and forests that act as natural carbon sinks.

– *Public awareness and education*: raising awareness of the impact of individual actions on climate change and promoting sustainable lifestyles. Encouraging local communities to participate in climate action and adaptation efforts.

5.5. Desertification

Desertification refers to the process of land deterioration in semi-arid and dry regions, resulting in the transformation of fertile land into a desert-like condition. It is caused by various factors such as climate change, overgrazing, deforestation and unsustainable agricultural practices. Desertification is the destruction of the biological potential of the soil, which can ultimately lead to desert like conditions. About 3.2 lakh km of the four states of Rajasthan, Punjab, Haryana and Gujarat fall in desert environments, largely man-made. It is spreading as a result of the exploitation of limited natural resources by over-grazing pastures, cutting of forests and poor cultivation-all contributing to wind erosion. The soil loses its productivity as a result of land degradation. The process involves the soil's ability to regenerate itself is greatly reduced or eliminated in the absence of elements that sustain plants. Therefore, the process that turns fertile soil into infertile land is known as desertification. Desertification leads to severe effects, including the loss of biodiversity, lowered the food production, increased poverty and forced migration. It affects millions of people globally, particularly those living in susceptible areas. Efforts to combat desertification involve sustainable land management practices, reforestation, water conservation and community participation. International agreements like the United Nations Convention to Combat Desertification aim to address this global issue and promote sustainable development. This phenomenon turns fertile land into desert land, typically as a result of deforestation, unsustainable agricultural practices, over-grazing and climate change. Desertification leads to the loss of vegetation cover and biodiversity, soil erosion and the depletion of water resources. It often occurs in arid, semi-arid and dry sub-humid regions, but can also affect other areas under certain conditions. Desertification is the process by which the biological productivity of arid and semi-arid lands is decreased due to external or human factors. Climate change, deforestation, overgrazing, poverty, unstable political systems, unsustainable irrigation techniques and/or a mix of these issues may all contribute to productivity declines. The notion pertains to the different mechanisms endangering all dryland ecosystems, encompassing deserts, grasslands and scrublands, rather than the actual growth of deserts. Some of the most prominent desertification-prone areas include the Sahel Region of Africa, the Gobi Desert of Asia and the Great Plains of North America. Desertification has wide-ranging effects on the environment, economy and society, making it a significant global concern that requires sustainable land management and conservation efforts. By implementing innovative and effective solutions in a coordinated and sustained manner, it is possible to address the challenges of desertification and encourage the wise use of natural resources and land for the betterment of both the present and future generations.

5.5.1. *Causes of desertification*

Desertification usually occurs in arid areas caused due to variety of factors, such as climate change and the overexploitation of soil through anthropogenic activities. Various causes of desertification are as follows:

– *Overgrazing*: if there are too many animals overgrazing in certain spots, it is difficult for plants to grow back. The animals hunt in the ecosystem, causing it to lose its previous green splendor.

– *Deforestation*: excess wood extraction and infrastructure expansion such as road building and urbanization contribute to the problems related to desertification.

– *Farming practice*: some farmers lack the knowledge necessary to manage the land well. They might practically clear the land of everything before relocating it to a new location. By stripping the soil of its nutrients, desertification becomes more of a reality for the area used for farming.

– *Excessive use of fertilizers and pesticides*: excessive amounts of fertilizer and pesticides to maximize crop yields in the short term often lead to significant damage for the soil, making it unsuitable for farming in the future.

– *Over-drafting of groundwater*: over-drafting is a process in which groundwater is extracted more than the equilibrium yield of the aquifer that is pumping or the excessive pulling up of groundwater from underground aquifers. Its depletion causes desertification.

– *Climate change*: as the temperatures rise, the periods of drought become more frequent. Desertification is becoming more and more prevalent. Unless climate change is slowed down, the huge areas of land will turn into a desert.

– *Population pressure and land degradation*: increasing population pressure, particularly in arid and semi-arid regions, can lead to unsustainable land use practices, land fragmentation and the conversion of natural landscapes into agricultural or urban areas, exacerbating the risk of desertification.

5.5.2. *Effects of desertification*

The process of desertification turns fertile land into desert and has far-reaching effects on the environment, economy and people living in affected areas. The effects of desertification are as follows:

– *Environmental impacts*: desertification leads to the loss of plant and animal species, disrupting ecosystems and reducing overall biodiversity. The topsoil becomes more vulnerable to erosion, leading to decreased soil fertility and it therefore becomes impossible to grow sustainable crops there. In some cases, desertification can result in the depletion of vegetation. It leads to the loss of fertile soil, reducing the land's capacity for agriculture and other forms of land use further exacerbating the environmental impact.

– *Biodiversity loss*: desertification contributes to habitat destruction and loss of biodiversity affecting various plant and animal species. Many species will not be able to adjust to the altered environmental conditions, which may lead to the extinction of certain species and serious population decline.

– *Economic impacts*: desertification reduces the productivity of agricultural land, leading to lower crop yields and food shortages. Communities relying on agriculture and livestock farming may face unemployment and poverty as their means of livelihood are threatened.

– *Social impacts*: desertification can force people to migrate in search of arable land and water, leading to displacement and potential conflicts over resources. Reduced access to food and water can lead to malnutrition and health issues in affected communities. The strain on resources and livelihoods can lead to social unrest and conflict within communities and between different social groups. In addition, scarcity of water and arable land, leading to competition and conflicts over these resources.

– *Impacts on food security and water scarcity*: reduced agricultural productivity due to soil degradation can increase food security, especially in regions where subsistence farming is prevalent. In addition, desertification also leads to reduced water availability as vegetation cover decreases and soil loses its capacity to retain moisture. This can exacerbate water scarcity issues in affected regions.

5.5.3. *Solution to desertification*

A holistic strategy including community involvement, environmental initiatives and sustainable land management is needed to combat desertification. Several strategies can be used to lessen and even reverse the consequences of desertification.

– *Afforestation*: planting trees and other vegetation in areas that have been subject to deforestation. This will stabilize the soil, reduce erosion and restore the biodiversity in degraded regions. They slow down global warming and contribute to maintaining a natural balance. Planting trees in the affected regions will prevent desertification and other environmental issues.

– *Implementing agroforestry practices*: such as intercropping trees with crops, helps improve soil fertility, reduce erosion and provide additional sources of income for communities.

– *Technology advances*: introducing the latest water-saving technologies, such as drip irrigation and rainwater harvesting, can help conserve water resources and support agriculture in arid regions. Encouraging the use of sustainable farming techniques, such as organic farming, crop rotation and conservation tillage, can help maintain soil health and productivity.

– *Restricting mining practices*: mining processes cause the destruction of large land areas. Therefore, it should be regulated by governments to keep the nature reserves intact and protect the natural habitats of many animal and plant species to a certain extent.

– *Policy interventions*: implementing policies and regulations that promote sustainable land use, discourage deforestation and incentivize conservation efforts can play a crucial role in combating desertification.

– *Education*: involving local communities in land management decisions and providing education on sustainable practices can foster a sense of ownership and promote long-term conservation efforts. In developing countries, education is an important tool that needs to be utilized to help people understand better ways to utilize the land. Education and awareness on sustainable practices should be provided to prevent land from turning into a desert.

5.6. Acid rain

Acid rain is a form of precipitation that contains high levels of sulfuric and nitric acids. It is primarily caused by emissions of sulfur dioxide (SO_2) and nitrogen oxides (NOx) from industrial processes, vehicle emissions and the burning of fossil fuels which react with the water molecules in the atmosphere to produce acids. "Acid rain" or "acid deposition" refers to any kind of precipitation that falls on Earth from the atmosphere and contains acidic elements, such as sulfuric or nitric acid, whether it's dry or wet. This can include hail, rain, snow, fog and acidic dust, among other things.

This phenomenon can have several environmental and societal impacts. It may be detrimental to aquatic life, vegetation and infrastructure. Since the 1970s, several countries have worked to limit the emissions of nitrogen and sulfur dioxide. Transitioning to cleaner energy sources such as renewable energy and natural gas can reduce the emission of pollutants contributing to acid rain. By addressing the root causes of acid rain and implementing measures to reduce pollutant emissions, it is possible to mitigate its harmful effects on the environment, infrastructure and human health.

5.6.1. *Sources of acid rain*

The primary sources of sulfur dioxide and nitrogen oxide emissions that lead to acid rain include the following:

– All bodies of water have some acid, but the issue with acid rain is that excessive amounts of acidic chemical compounds, including nitric and sulfuric acid, are created when sulfur dioxide (SO_2) and nitrogen oxide

(NOx) react with water and atmospheric oxygen to lower the natural pH of water.

– Natural processes like decomposing plants and biological activities, in addition to natural calamities like wildfires, lightning and volcanic eruptions that shoot pollutants into the air, are natural sources of gasses that cause acid rain.

– However, man-made sources include emissions from industries and power plants that use coal and oil as fossil fuels. The three main sources of chemical gases are power plants, oil refineries and cars. The primary source of gaseous emissions is the burning of coal and other fossil fuels in electricity-generating facilities. About 25% of NOx and 60% of SO_2 in the atmosphere is caused by them.

– Moreover, substantial concentrations of gaseous pollutants are released into the atmosphere by the exhaust of cars, trucks and buses, particularly in populated areas with considerable traffic.

– These complex pollutants can travel great distances on wind- or jet-borne routes, where they react with oxygen and water to form acid. Rain can have its natural pH lowered by this acid. The more the acidity, the lower the pH value.

5.6.2. *Effects of acid rain*

The impact of acid rain is widespread affecting the environment, human health and infrastructure. Some of the key effects include the following:

– When acid rain seeps into the soil, it dissolves vital elements like calcium and magnesium that are necessary for trees to develop properly. In the absence of these vital nutrients, the trees and plants are less healthy and more vulnerable to infections and damage by cold weather and insects. In addition, acid rain releases aluminum into the soil, which hinders trees' ability to absorb water and prevents the growth of trees.

– Acid rain damages forests, lakes and aquatic ecosystems by altering the pH levels of soil and water, which can harm plant and animal life. The greatest impact on water bodies and the aquatic ecosystem comes from direct acid rain or precipitation that travels through fields, forests and roads and into streams and lakes. This precipitation typically contains some aluminium that has leached from the soil.

– It lowers the PH levels in most water bodies like lakes and rivers. At lower pH levels, most aquatic species cannot survive and sometimes may die. Increased levels of acid and aluminum in water bodies are toxic to aquatic life and may also be deadly.

– The pollutants that cause acid rain can contribute to respiratory problems in humans, particularly for those with preexisting conditions such as asthma. Acid rain water causes serious health problems for human beings. Inhalation of air pollutants associated with acid rain can lead to respiratory problems and other health issues including asthma and chronic bronchitis as well as an increase in heart disease risks.

– Not only are living creatures affected, but acid rain causes corrosion of buildings and structures, and accelerates the deterioration of buildings, monuments and infrastructure made of stone, metal and other materials. Historical and cultural sites can be significantly impacted by the corrosive effects of acid rain, leading to irreversible damage. It also leads to corrosion of various types of metals.

– The effects of acid rain are not limited to the areas where the pollutants are produced. Due to the long-range transport of pollutants, acid rain can affect regions far from its sources. This global impact underscores the need for international cooperation to address the issue.

5.6.3. *Solution to control acid rain*

Acid rain can be stopped in several ways. The government also plays a significant role in restricting the use of fossil fuels and concentrating on more sustainable energy sources like solar, wind and water energy. Addressing the problem of acid rain requires a comprehensive approach at local, national and international levels. Some potential solutions include:

– *Reduction of emissions*: implementing technologies and regulations to reduce the release of sulfur dioxide and nitrogen oxides from industrial processes and power generation can help mitigate acid rain. The biggest step to prevent acid rain is to conserve energy by simply shutting off electrical appliances.

– *Power plants*: these must also contribute in some way. Using low-sulfur coal or washing coal to remove the parts of the sulfur are two things they can do.

Additionally, they have access to scrubber devices. They can extract sulfur dioxide from the smokestack's outgoing gasses.

– *Limiting emissions*: implementing and enforcing regulations to limit sulfur dioxide and nitrogen oxide emissions from industrial facilities and vehicles. In addition, promoting the use of clean and renewable energy sources as alternatives to fossil fuels to reduce emissions.

– *Technology*: developing and deploying technologies such as flue-gas desulfurization systems in power plants to remove sulfur dioxide from emissions.

5.7. Ocean acidification

Ocean acidification is a significant environmental concern that has received more attention in recent years. This process refers to the ongoing decrease in the pH of the Earth's oceans, primarily caused by the uptake of carbon dioxide (CO_2) from the atmosphere. The subsequent chemical reactions result in a reduction of the seawater's pH, leading to a range of ecological, economic and societal consequences. Ocean acidification is primarily caused by the increased concentration of carbon dioxide in the atmosphere due to human activities such as the burning of fossil fuels and deforestation. Additionally, the primary cause of ocean acidification is the increased concentration of CO_2 in the Earth's atmosphere, largely due to human activities such as the burning of fossil fuels, deforestation and industrial processes. The ocean absorbs approximately 30% of the CO_2 emitted into the atmosphere, leading to the formation of carbonic acid and subsequent changes in the ocean's chemistry. Other factors such as nutrient runoff from land, which leads to algal blooms and the release of sulfur and nitrogen oxides from industrial processes, can also contribute to localized instances of acidification. Ocean acidification represents a complex and multifaceted challenge with far-reaching implications for marine life, economies and societies. Addressing this issue requires concerted efforts at both local and global levels, involving a combination of emissions reductions, ecosystem conservation and ongoing scientific research to better understand and mitigate the impacts of this phenomenon. Stakeholders across various sectors must work collaboratively to address ocean acidification and safeguard the health and resilience of our oceans for future generations.

5.7.1. *Impact of ocean acidification*

5.7.1.1. *Effects on marine life*

The acidification of the oceans has far-reaching effects on marine life. It impacts the ability of shell-forming organisms, such as corals, Molluscs and some types of plankton, to build and maintain their shells. Additionally, it can disrupt the marine food web and lead to detrimental effects on fish populations and overall biodiversity. Ocean acidification disrupts marine ecosystems by affecting the behavior and survival of various species. It can also impact the biogeochemical cycling of elements and ultimately influence the productivity and stability of marine ecosystems. Ocean acidification poses a significant threat to marine organisms that rely on calcium carbonate to build their shells and skeletons, such as corals, molluscs and certain types of plankton. The decreased availability of carbonate ions in more acidic waters makes it harder for these organisms to maintain their calcium carbonate structures, leading to weaker shells and reduced growth rates. Disruptions to calcifying organisms can have cascading effects throughout marine food webs and ecosystems, impacting the productivity and biodiversity of marine environments. Ocean acidification can also affect fish populations and the fishing industry, potentially leading to reduced fish stocks and economic impacts on communities that depend on marine resources for their livelihoods.

5.7.1.2. *Societal and economic implications*

The potential decline in fish populations due to ocean acidification could have far-reaching implications for global food security, especially for communities that rely heavily on seafood as a primary protein source. In addition, the economic costs associated with ocean acidification are substantial, encompassing losses in fisheries, tourism (e.g. due to coral bleaching) and potential impacts on other industries linked to marine ecosystems.

5.7.2. *Mitigation and adaptation strategies*

– *CO_2 emissions reduction*: addressing the root cause of ocean acidification involves reducing CO_2 emissions through measures such as transitioning to renewable energy sources, improving energy efficiency and implementing carbon capture and storage technologies.

– *Marine protected areas*: establishing and effectively managing marine protected areas can help conserve and protect vulnerable marine ecosystems from additional stressors, including ocean acidification.

– *Research and monitoring*: further research and monitoring efforts are crucial for better understanding the specific impacts of ocean acidification on different marine organisms and ecosystems, as well as for developing targeted adaptation strategies.

– *Educational outreach and awareness*: promoting public awareness and educating communities about ocean acidification is crucial. It involves fostering an understanding of the problem, its consequences and the actions needed to address this ecological threat. Educational programs can empower individuals to contribute to conservation efforts and advocate for sustainable practices.

5.8. Urban waterlogging

Urban waterlogging occurs when heavy and intense rainfall for a short time continuously exceeds the capacity of the urban drainage system. It is also defined as the phenomenon when the runoff yield volume exceeds the carrying capacity of the urban drainage system under the condition of high-intensity and short-duration rainstorms. It causes rainwater drainage to become less effective, which in turn causes large-scale water accumulation on the urban surface. Waterlogging is a fundamental problem during moderate and extreme rain weather for low-lying areas. It is one of the major problems of the urban areas, due to climate change, uncontrollable growth of population, unplanned urbanization and poor drainage system, creating panic for local people as well as the administration. The severity of urban waterlogging is increasing due to climate change and increased urbanization. It occurs when the natural absorption of water is obstructed by man-made structures in urban environments. This leads to water accumulating on the surface, causing damage to infrastructure. The causes of urban flooding can vary from inadequate drainage systems to climate change-induced heavy rainfall events. Rainstorm-induced urban waterlogging is one of the most common, hazardous and catastrophic natural disasters in the world particularly in regions that experience heavy rainfall leading to inundation of areas with water, either due to excessive rainfall, poor drainage infrastructure, or rising groundwater levels and unsustainable land management practices. It can have significant environmental, social and economic impacts, affecting both urban and rural areas.

With a changing climate and the rapid growth of urban areas, the instances of urban flooding are becoming increasingly severe and complex issues predicted to pose significant threats to both lives and property. Urban regions have been suffering from waterlogging problems since ancient times, and this hazard is one of the problematic matters for these areas. Countries like India, Bangladesh and Pakistan face significant waterlogging issues, particularly in densely populated urban areas and low-lying agricultural regions. Rapid urbanization, inadequate drainage systems and heavy monsoon rains contribute to waterlogging in cities like Mumbai, Dhaka and Karachi. Bangladesh is experiencing rapid urbanization and lacks adequate drainage systems to manage heavy rainfall and stormwater runoff. Climate change can exacerbate urban waterlogging by overwhelming existing drainage systems and increasing the risk of flooding in low-lying areas. In addition, existing drainage networks are often outdated, undersized or poorly maintained, leading to waterlogging during monsoon seasons. This rapid growth results in unplanned development, including the construction of buildings, roads and other infrastructure without proper consideration of drainage needs. As a result, natural drainage pathways are often blocked or covered, exacerbating waterlogging issues. Southeast Asian countries such as Indonesia, the Philippines and Vietnam also experience waterlogging, especially in coastal areas prone to flooding during typhoon and monsoon seasons. Inadequate drainage infrastructure and land subsidence exacerbate waterlogging issues in cities like Jakarta and Manila. Urban flooding has turned into a significant worry for networks living in urban regions. As per the study, 55% of the world's population lives in urban areas, a proportion that is expected to increase to 68% by 2050 (UN World Urbanization Prospectus, 2018). Cities like China, Delhi, Kolkata, Beijing, Mumbai, Tamil Nadu, Bihar and Bangalore are among the quickest developing urban communities at the national and international levels that experience traffic disruptions due to waterlogging during heavy rains. Urban flooding and waterlogging normally create havoc and panic circumstances for the individuals living in urban areas due to heavy rain. High-intensity precipitation and low percolation results flood in urban areas, which are expanding each year. Urban flooding is a significant issue of disaster management and there is an immediate need to concentrate on the identification and evaluation of vulnerable areas. Flood hazard mapping, mitigation plans, early warnings and forecasting will help in proper decision-making toward diminishing and management of urban floods. Encroachment of natural drainages and lack of proper outlets for subways, wetlands, floodplains, etc., obstructing the flood waters, particularly in low-lying urban areas. It requires a multidisciplinary approach to address these challenges, posed by urban flooding. Studying waterlogging hazards is crucial for mitigating the negative impacts on the environments and infrastructure. It is vital to develop effective strategies to manage and reduce the risks associated with urban flooding and waterlogging. If the problem of waterlogging in one area can be resolved, the same process may be used to reduce the issue in other areas as well.

5.8.1. *Causes of waterlogging*

Driving factors for urban/rural waterlogging are as follows:

1) *Precipitation*: heavy rainfall is one of the main reasons for urban waterlogging. Intense or prolonged periods of rainfall can lead to an increase in surface runoff. When the rate of rainfall exceeds the infiltration capacity of the soil (its ability to absorb water), excess water accumulates on the surface, potentially causing flooding. In addition, changes in weather patterns due to climate change can lead to more intense and frequent rainfall events, which inundate certain regions in the district for several days mainly due to the drainage congestion and results in urban flooding.

2) *Elevation*: regions with lower elevations are more prone to flooding because water naturally flows downhill. When heavy rainfall or snowmelt occurs, water accumulates in low-lying areas, resulting in flooding. Cities built in low-lying areas or areas with poor drainage are more susceptible to flooding, as water naturally collects in these locations. Elevation is commonly considered a topographical factor in urban flooding.

3) *Infrastructure*: inadequate or poorly maintained drainage systems, including stormwater drains, sewers and culverts, can lead to water accumulation on streets and in low-lying areas during heavy rainfall. Urban flooding is greatly influenced by the drainage infrastructure's functionality. Faulty design, solid waste, rubbish dumping, encroachment and unauthorized structures, siltation, lack of renovation and re-excavation are the leading causes of drainage congestion.

4) *Urbanization*: the expansion of urban areas often involves replacing natural surfaces with increased impervious. Surfaces like roads, concrete

sidewalks parking lots and other concrete structures lead to a decrease in rainfall infiltration and an increase in surface runoff. This reduces the land's ability to absorb water, increasing runoff and the likelihood of flooding.

5) *Topography*: this includes the nature of slope and elevation. Cities built in low-lying areas or areas with poor natural drainage are more susceptible to flooding, as water naturally collects in these locations.

6) *Lack of urban planning*: inadequate urban planning, including improper zoning regulations and building codes, can lead to construction in flood-prone areas or inadequate flood water management measures. The primary cause of the non-implementation of the district plan is lack of financial resources.

5.8.2. *Impact of urban waterlogging*

The consequences of urban flooding can be devastating, resulting in the loss of human life and extensive property damage. Waterlogging is said to be the major problem causing devastating and long-term effects in most of the urban areas. It can disrupt economic activities and livelihoods. In urban areas, flooded roads, bridges and public transportation systems can hinder mobility, commerce and access to employment opportunities, affecting local economies and employment rates. It can cause significant issues for cities, including damage to infrastructure, transportation disruptions and public health concerns including injuries, vector-borne illnesses and waterborne infections. The spread of germs, pollutants and harmful substances by contaminated floodwaters raises the danger of respiratory ailments, skin infections, gastrointestinal infections and other health issues. Damage to critical infrastructure, including utilities, roads, bridges, schools and communication networks, can sustain damage from waterlogging and flooding. Damage to infrastructure can interfere with the provision of basic services like electricity, water, sanitation and healthcare, which can have an impact on the well-being and quality of life. It takes a lot of money and time to repair and restore damaged infrastructure. Disaster due to urban waterlogging is a key problem that restricts the development of urban ecology in particular areas causing problems in the transport system and overall economic disruption during the rainy season. During monsoon season, waterlogging due to flooding occurs frequently and creates plenty of issues for the occupants of this region. It hampers their monetary exercises in everyday life, the flow of vehicular traffic, etc. Even a moderate to heavy downpour will submerge or inundate various parts of the area such as lanes, highways, housing societies, shops, and markets. Waterlogging during flooding has the potential to affect community networks and social cohesiveness. Children, the elderly, those with disabilities and disadvantaged populations are examples of vulnerable groups who may be disproportionately impacted by this hazard and may need special attention and assistance. Nowadays, more attention needs to be paid toward waterlogging and flood hazards. Deep understanding and utilization of new and innovative methods should be used to address the problem and to ensure sustainable development.

5.8.3. *Strategies to prevent waterlogging in urban areas*

Researchers and engineers are exploring innovative modeling techniques and methods to better understand and mitigate the impacts of urban flooding to create a more resilient and sustainable urban environment for future generations.

– *Improved infrastructural systems*: stormwater drainage systems play a crucial role in efficiently managing excess rainwater in urban areas. Regular desilting of drains and culverts is essential to ensure they remain unblocked and can effectively function during heavy rainfall. Additionally, incorporating green infrastructure such as permeable pavements and parking lots allows rainwater to infiltrate into the ground, reducing surface runoff. Implementing green roofs on buildings further helps to absorb rainwater and decrease the volume of water entering the drainage system.

– *Flood control measures*: these measures are also vital in preventing water-related disasters. Constructing retention ponds or basins provides temporary storage for excess rainwater, preventing the drainage system from becoming overwhelmed. Installing flood barriers or gates in low-lying areas helps to keep water out of urban spaces during heavy rainfall or flooding.

– *Urban planning and land use strategies*: to prevent waterlogging disasters, urban planning and land use strategies are crucial. This involves the implementation of zoning regulations to restrict construction in flood-prone areas and promote the use of open spaces for water absorption. By incorporating water management into urban planning, new developments can consider waterlogging risks and mitigate potential disasters.

– *Public awareness and early warning systems*: establishing public awareness and early warning systems. These systems serve to alert residents and authorities about impending heavy rainfall and potential waterlogging. Educating the public about the importance of proper waste disposal and the impact of urban activities on waterlogging can help in disaster prevention.

– *Maintenance and monitoring for disaster prevention*: regular maintenance of drainage systems, pumps and other infrastructure is essential to prevent malfunctions during heavy rainfall and avoid waterlogging disasters. Implementing monitoring systems to track water levels, rainfall patterns and drainage system performance can lead to an early detection of potential waterlogging issues. Additionally, implementing and promoting sustainable development practices and resilient water systems can prevent such disasters.

5.9. Questions

1) What is global warming and how does it impact the environment?

2) Briefly describe primary factors of climate change and its impact.

3) Explain the concept of soil degradation and its implications for agricultural productivity and ecosystem health.

4) What is the ozone layer, and how does it protect life on Earth from harmful ultraviolet radiation?

5) What are the main causes of ozone depletion, and how do human activities such as the use of chlorofluorocarbons (CFCs) contribute to this phenomenon?

6) Discuss the main causes of soil degradation and potential measures to improve soil quality and fertility.

7) What are the primary causes of desertification, and how do human activities contribute to this process?

8) Describe the concept of plastic pollution in marine environments.

9) Discuss the sources and pathways of plastic pollution, its impacts on marine ecosystems and wildlife, and potential solutions to address the issue.

10) How has climate change impacted global weather patterns and ecosystems over the past century?

11) What are some sustainable land management practices that can help prevent or reverse desertification?

12) Describe the effects of deforestation on biodiversity loss and ecosystem services in tropical rainforests.

13) How do plastic pollution and microplastics affect marine environments and aquatic species?

14) What are the primary causes and consequences of air pollution in urban areas around the world?

15) Discuss the role of industrial agriculture in contributing to soil degradation and loss of arable land.

16) How are rising sea levels threatening coastal communities and low-lying island nations?

17) Describe the social, economic and environmental implications of water scarcity in arid regions and drought-prone areas.

18) Describe the processes and causes of land degradation, and explain its implications for agriculture and food production.

19) Explore methods for restoring degraded land and promoting sustainable land management practices.

20) How has plastic pollution become a global environmental challenge and what are its effects on marine life?

21) Explore potential solutions to reduce and manage plastic waste at local and global levels.

22) Describe how global warming is altering weather patterns and leading to more frequent and severe extreme weather events.

23) Explore the connections between rising temperatures, changes in precipitation patterns and the frequency of heat waves and droughts.

24) What are the main chemical components of acid rain, and how do they form in the atmosphere?

25) How does desertification exacerbate issues related to food security and water scarcity in affected regions?

26) How does acid rain impact aquatic ecosystems and what are the long-term consequences for freshwater organisms?

27) What are some measures that can be taken to mitigate the effects of acid rain on soil quality and agricultural productivity?

28) How does acid rain contribute to the deterioration of historical monuments and buildings made of limestone or marble?

29) What are the primary causes of urban waterlogging, and how do they impact the infrastructure and daily life in urban areas?

30) How can local communities and residents contribute to preventing and managing urban waterlogging?

31) Which processes contribute to ocean acidification, and how do they alter the chemical composition of seawater?

32) What are the potential global implications of ocean acidification on fisheries, food security, and the overall health of the oceans?

33) Which strategies and technologies are being developed to mitigate the effects of ocean acidification, and how can societies adapt to these changes?

34) What are the potential health impacts of global warming on human populations, including the spread of diseases, changes in food production and the frequency of extreme weather events?

35) What are some effective strategies that governments, businesses and individuals can implement to mitigate the effects of global warming and adapt to a changing climate?

36) How does climate change impact marginalized and vulnerable groups?

37) What are the primary sources of greenhouse gas emissions, and how do different human activities contribute to the accumulation of these gases in the atmosphere?

38) What strategies are being pursued to reduce greenhouse gas emissions?

39) What are the primary factors contributing to waterlogging in urban areas?

40) Discuss the importance of community engagement and stakeholder involvement in addressing urban waterlogging.

5.10. Multiple-choice questions (MCQs)

1) Which human activity contributes to the formation of acid rain?

a) **Industrial emissions**

b) Oil spills

c) Deforestation

d) Agricultural runoff

2) What is a major cause of land pollution from human activities?

a) Volcanic eruptions

b) Deforestation

c) Nuclear accidents

d) **Improper waste disposal**

3) What is the primary cause of soil contamination from industrial activities?

a) Oil spills

b) **Pesticide runoff**

c) Nuclear accidents

d) Deforestation

4) Which human activity contributes to the depletion of the ozone layer?

a) Deforestation

b) **Use of aerosol sprays**

c) Agricultural practices

d) Mining

5) Man-made disasters include:

a) Wars

b) Terrorism

c) Industrial hazards

d) **All of the above**

6) What is it called when a large number of people in a community contract a disease at the same time?

a) Influx

b) Black death

c) **Epidemic**

d) Parasite

7) The Bhopal Gas Tragedy occurred on:

a) **December 3, 1984**

b) December 5, 1985

c) November 3, 1984

d) December 3, 1987

8) Typhoid-causing bacteria are:

a) Shigella

b) **Salmonella**

c) HIV

d) HCV

9) Which metal causes Minamata disease?

a) Iron

b) **Mercury**

c) Gold

d) Silver

10) The toxic gas released from the Bhopal gas tragedy was:

a) **Methyl isocyanate**

b) Potassium isocyanate

c) Sodium isocyanate

d) Methyl isothiocyanate

11) IPCC stands for?

a) **Intergovernmental Panel on Climate Change**

b) Intergovernmental Panel on Climatic Condition

c) Intergovernmental Programme on Climate Change

d) International Panel on Climate Change

12) Which of the following emits dangerous radioactive wastes forcing the government to eliminate its use?

a) **Nuclear energy**

b) Water energy

c) Geothermal energy

d) Solar energy

13) Which of the following is related to acid rain?

a) Carbon dioxide

b) Nitrogen

c) **Sulfur dioxide**

d) Carbon monoxide

14) Which gas caused the Bhopal gas tragedy?

a) **Methyl isocyanate**

b) Carbon dioxide

c) Oxygen

d) Methane

15) What is the primary greenhouse gas emitted through human activities?

a) **Carbon dioxide**

b) Methane

c) Nitrous oxide

d) Sulfur

16) Which of the following is a major cause of deforestation?

a) Urbanization

b) **Agriculture**

c) Mining

d) Industrialization

17) What is the largest source of water pollution in many countries?

a) Industrial discharges

b) **Agricultural runoff**

c) Oil spills

d) Chemicals

18) Which of the following is a nonrenewable energy source?

a) Solar power

b) Wind energy

c) **Coal**

d) Tidal energy

19) What is the primary cause of soil erosion?

a) **Deforestation**

b) Urbanization

c) Mining

d) Groundwater extraction

20) What is the main factor contributing to the depletion of the ozone layer?

a) Carbon dioxide emissions

b) **Chlorofluorocarbons (CFCs)**

c) Methane emissions

d) Sulfur emissions

21) What is the leading cause of species extinction worldwide?

a) Climate change

b) **Habitat destruction**

c) Pollution

d) Global warming

22) What is the primary source of indoor air pollution in many developing countries?

a) Tobacco smoke

b) **Cooking with solid fuels**

c) Radon gas

d) Smog

23) What is a significant contributor to ocean acidification?

a) **Industrial emissions**

b) Deforestation

c) Carbon dioxide absorption

d) Decrease in oxygen

24) What is a major driver of desertification?

a) **Overgrazing**

b) Reforestation

c) Urbanization

d) Industrialization

25) What is a common by-product of mining activities that can lead to water pollution?

a) Sulfur dioxide

b) **Heavy metals**

c) Carbon monoxide

d) Minerals

26) What is a significant cause of coral reef degradation?

a) Overfishing

b) **Ocean acidification**

c) Noise pollution

d) Air pollution

27) What is a primary factor contributing to the loss of biodiversity?

a) Urban greening

b) **Invasive species**

c) Reforestation

d) Overgrazing

28) What is a major cause of river pollution in urban areas?

a) Agricultural runoff

b) **Industrial discharges**

c) Oil spills

d) Urbanization

29) What is a significant driver of air pollution in megacities?

a) Deforestation

b) **Vehicle Emissions**

c) Mining activities

d) Soil pollution

30) What is a common consequence of eutrophication in lakes and rivers?

a) Increased oxygen levels

b) **Algal blooms**

c) Reduced nutrient levels

d) Climate change

References

Adikari, Y. and Yoshitani, J. (2009). *Global Trends in Water-Related Disasters: An Insight for Policy Makers*. International Centre for Water Hazard and Risk Management, Tsukuba, and UNESCO World Water Assessment Programme, Perugia.

Agrawal, S. and Srivastava, N.B. (translators) (1986). *Prakritik Prokop*. Joint Assistance Centre, Gurgaon.

Agrawal, R., Singh, S.K., Kanga, S., Sajan, B., Meraj, G., Kumar, P. (2024). Advancing flood risk assessment through integrated hazard mapping: A Google Earth Engine-based approach for comprehensive scientific analysis and decision support. *Journal of Climate Change*, 10(1), 47–60.

Baas, S. (2010). *Disaster Risk Management Systems Analysis: A Guide Book: No. 13 (Environment and Natural Resources Management Series)*, 1st edition. Food & Agriculture Organization of the United Nations (FAO), Rome.

Bawa, D.S. (1980). *New Strategy for Industrial Development of Flood and Drought Prone Areas*. The University of Michigan, Bawa.

Brody, S.D., Highfield, W.E., Kang, J.E. (2011). *Rising Waters: The Causes and Consequences of Flooding in the United States*. Cambridge University Press, Cambridge.

Carter, W.N. (1991). *Disaster Management: A Disaster Manager's Handbook*. Asian Development Bank, Manila.

Carter, W.N. (2008). *Disaster Management: A Disaster Manager's Handbook*. Asian Development Bank, Manila.

Chand, K., Kumar, A., Meraj, G., Thakur, T., Farooq, M., Kumar, P., Singh, S.K., Kanga, S., Debnath, J. (2024). Integrating geoenvironmental and socioenvironmental analyses for flood vulnerability assessment in the Kullu Valley, Himachal Pradesh, India. *International Journal of Disaster Risk Reduction*, 108, 104494.

Coppola, D. (2015). *Introduction to International Disaster Management*, 3rd edition. Butterworth Heinemann, Oxford [Online]. Available at: https://www.researchgate.net.

Debnath, J., Sahariah, D., Mazumdar, M., Lahon, D., Meraj, G., Hashimoto, S., Kumar, P., Singh, S.K., Kanga, S., Chand, K. et al. (2023). Evaluating flood susceptibility in the Brahmaputra River basin: An insight into Asia's Eastern Himalayan floodplains using machine learning and multi-criteria decision-making. *Earth Systems and Environment*, 7, 733–760. doi: 10.1007/s41748-023-00358-w.

Debnath, J., Sahariah, D., Nath, N., Saikia, A., Lahon, D., Islam, Md.N., Hashimoto, S., Meraj, G., Kumar, P., Singh, S.K. et al. (2024a). Modelling on assessment of flood risk susceptibility at the Jia Bharali River basin in Eastern Himalayas by integrating multicollinearity tests and geospatial techniques. *Modeling Earth Systems and Environment*, 10, 2393–2419. doi: 10.1007/s40808-023-01912-1.

Debnath, J., Debbarma, J., Debnath, A., Meraj, G., Chand, K., Singh, S.K., Kanga, K., Kumar, P., Sahariah, D., Saikia, A. (2024b). Flood susceptibility assessment of the Agartala Urban Watershed, India, using machine learning algorithm. *Environmental Monitoring and Assessment*, 196, 110. doi: 10.1007/s10661-023-12240-3.

Diwan, P. (2010). *A Manual on Disaster Management*. Pentagon Press, New Delhi.

Drake, F. (2000). *Global Warming: The Science of Climate Change*. Oxford University Press, New York.

Edwards, B. (2005). *Natural Hazards*. Cambridge University Press, Cambridge.

Farooq, M., Rashid, H., Meraj, G., Kanga, S., Singh, S.K. (2022). Assessing the microclimatic environmental indicators of climate change of a temperate valley in the Western Himalayan region. In *Climate Change, Disaster and Adaptations*, Siddiqui, A.R. and Sahay, A. (eds). Springer, Cham. doi: 10.1007/978-3-030-91010-5_5.

Government of India (1997). *Vulnerability Atlas of India*. Government of India, India.

Gupta, H.K. (2003). *Disaster Management*. Universities Press, Hyderabad.

Gupta, S.K., Kanga, S., Meraj, G., Pankaj, K., Singh, S.K. (2023). Uncovering the hydro-meteorological drivers responsible for forest fires utilizing geospatial techniques. *Theoretical and Applied Climatology*, 153, 675–695. doi: 10.1007/s00704-023-04497-y.

Hitouri, S., Mohajane, M., Lahsaini, M., Ali, S.A., Setargie, T.A., Tripathi, G., D'Antonio, P., Singh, S.K., Varasano, A. (2024). Flood susceptibility mapping using SAR data and machine learning algorithms in a small watershed in Northwestern Morocco. *Remote Sensing*, 16(5), 858. doi: 10.3390/rs16050858.

IGNOU (n.d.). Programme guide. IGNOU – School of Social Sciences, New Delhi [Online]. Available at: http://www.ignou.ac.in.

IPCC (2001). *Third Assessment Report of the Intergovernmental Panel on Climate Change*. Cambridge University Press, Cambridge [Online]. Available at: https://www.ipcc.ch.

Jain, N.K. (ed.) (1995). *Moving Technology*, volume 10. Council for Advancement of People's Action and Rural Technology (CAPART), New Delhi.

Joy, J., Kanga, S., Singh, S.K. (2019). Kerala flood 2018: Flood mapping by participatory GIS approach, Meloor Panchayat. *International Journal of Emerging Technologies*, 10(1), 197–205.

Kanga, S., Sadhu, S., Singh, S.K. (2017). Optimum path from tourist places to railway station and hospitals & fire services using GIS-based network analysis: A case study of Jaisalmer and Sam Blocks of Jaisalmer. *International Journal of Technology Research and Management*, 4(8), 1–5.

Kanga, S., Rather, M.A., Farooq, M., Singh, S.K. (2021). GIS-based forest fire vulnerability assessment and its validation using field and MODIS data: A case study of Bhaderwah Forest Division, Jammu and Kashmir (India). *Indian Forester*, 147(2), 120–136.

Kanga, S., Meraj, G., Farooq, M., Singh, S.K., Nathawat, M.S. (2022a). Disasters in the complex Himalayan terrains. In *Disaster Management in the Complex Himalayan Terrains*, Kanga, S., Meraj, G., Farooq, M., Singh, S.K., Nathawat, M.S. (eds). Springer, Cham. doi: 10.1007/978-3-030-89308-8_1.

Kanga, S., Meraj, G., Farooq, M., Singh, S.K., Nathawat, M.S. (eds) (2022b). *Disaster Management in the Complex Himalayan Terrains: Natural Hazard Management, Methodologies and Policy Implications*, 1st edition. Springer, Cham. doi: 10.1007/978-3-030-89308-8.

Khan, J.A. (2001). *Manual on Natural Disaster Management in India.* National Centre for Disaster Management, Ministry of Agriculture, Government of India, India.

Kotecha, M.J., Tripathi, G., Singh, S.K., Kanga, S. (2023). GIS-based novel ensemble MCDM-AHP modeling for flash flood susceptibility mapping of Luni River basin, Rajasthan. In *River Conservation and Water Resource Management*, Rai, P.K. (ed.). Springer, Cham. doi: 10.1007/978-981-99-2605-3_14.

Kumar, R. (2019). *Introduction to Disaster & Its Management*, 1st edition. 24by7 Publishing, North Dumdum.

Kumar, S., Singh, S.K., Kanga, S. (eds) (2024). *Disaster Management and Environmental Sustainability*. John Wiley & Sons, New York.

Lahon, D., Sahariah, D., Debnath, J., Nath, N., Meraj, G., Farooq, M., Kanga, S., Singh, S.K. (2023). Growth of water hyacinth biomass and its impact on the floristic composition of aquatic plants in a wetland ecosystem of the Brahmaputra floodplain of Assam, India. *Peer J*, 11, e14811.

Lino Grima, A.P. (1992). Book reviews: Palm, R.I. 1990: Natural hazards: An integrative framework for research and planning. Baltimore and London: The Johns Hopkins University Press. *Progress in Human Geography*, 16(1), 142–144. doi: 10.1177/030913259201600128.

McGraw-Hill (2005). *McGraw-Hill Concise Encyclopedia of Science & Technology*, 5th edition. McGraw-Hill Education, New York.

Misra, G.K. and Mathur, G.C. (eds) (1995). *Natural Disaster Reduction*. Reliance Publishing House and Indian Institute of Public Administration, New Delhi.

Pant, H., Verma, J., Surya, S. (2020). *Environmental Issues: Local, Regional and Global Environmental Issues*. Society of Biological Sciences and Rural Development, New Jhusi [Online]. Available at: https://www.researchgate.net/publication.

Pimm, S.L. and Rafferty, J.P. (n.d.). Desertification. In *Encyclopedia Britannica* [Online]. Available at: https://www.britannica.com.

Prakash, I. (1994). *Disaster Management*. Rashtra Prahari Prakashan Sahibabad, Ghaziabad.

Raju, N.V.S. (2021). *Disaster Management: A Comprehensive Approach*. BS Publications, Petersfield.

Rangachari, R. (1986). *Flood Forecasting and Warning Network on Interstate Rivers of India*. Central Water Commission, New Delhi.

Rangachari, R. (ed.) (1989). *Manual on Flood Forecasting*. Central Water Commission, New Delhi.

Rather, M.A., Meraj, G., Farooq, M., Shiekh, B.A., Kumar, P., Kanga, S., Singh, S.K., Sahu, N., Tiwari, S.P. (2022). Identifying the potential dam sites to avert the risk of catastrophic floods in the Jhelum Basin, Kashmir, NW Himalaya, India. *Remote Sensing*, 14(7), 1538. doi: 10.3390/rs14071538.

Razo, S.N. (n.d.). Hydro-meterological disasters. National Institute of Disaster Management, Ministry of Home Affairs, Government of India [Online]. Available at: https://nidm.gov.in/PDF/Disaster_hymet.pdf.

Sahni, P. and Ariyabnndu, M.M. (eds) (2003). *Disaster Risk Reduction in South Asia*. Prentice-Hall of India, New Delhi.

Sahoo, R.K. and Senapati, T. (2012). *Disasters: Management and Mitigation*. Dominant Publishers, New Delhi.

Salle, M.N.L. (2019). Hydro-meteorological hazards. SlideShare [Online]. Available at: https://www.slideshare.net/MaNestleLyzaSalle/hydrometeorological-hazards-179090107.

Sharrna, V.K. (ed.) (1995). *Disaster Management*. Indian Institute of Public Administration, New Delhi.

Sharma, S.C. (2018). *Disaster Management*. Khanna Publishing House, New Delhi.

Sharma, V.K., Kulkarni, A., Sharma, V., Jha, S.K., Kulshrestha, S.M., Sharma, V.K., Arora, S.B. (2017a). CDM-01 foundation course in disaster management. eGyanKosh, IGNOU Self Learning Material (SLM) 02. School of Social Sciences (SOSS), Certificate Programme in Disaster Management (CDM).

Sharma, V.K., Kulkarni, A., Sharma, V., Jha, S.K., Kulshrestha, S.M. (2017b). CDM-02 disaster management: Methods and techniques. eGyanKosh, IGNOU Self Learning Material (SLM) 02. School of Social Sciences (SOSS), Certificate Programme in Disaster Management (CDM) [Online]. Available at: https://www.egyankosh.ac.in.

Singh, A.K. (2008). *Handbook on Disaster Management*. Rajat Publications, Delhi.

Singh, S.K. (2017). Evaluation of flood and waterlogging dynamics in Indo-Gangetic plain using geospatial technique: A review. *Open Access International Journal of Science and Engineering*, 2(3), 1–7.

Singh, S.K., Pandey, A.C., Nathawat, M.S. (2011). Rainfall variability and spatiotemporal dynamics of flood inundation during the 2008 Kosi flood in Bihar State, India. *Asian Journal of Earth Sciences*, 4(1), 9–19.

Singh, S.K., Kanga, S., Tripathi, G. (2017a). Forest fire hazards vulnerability and risk assessment in Bhaji forest range of Himachal Pradesh (India): A geospatial approach. *Journal of Remote Sensing & GIS*, 8(1), 25–40.

Singh, S.K., Kanga, S., Kumar, S. (2017b). Climate-induced variation in forest fire using remote sensing and GIS in Bilaspur district of Himachal Pradesh. *International Journal of Engineering and Computer Science*, 6(6), 21695–21702.

Singh, S.K., Kanga, S., Đurin, B., Kranjčić, N., Chaurasia, R., Markovinović, D. (2021). Flood risk modeling using HEC-RAS and geospatial techniques. *E-ZBORNIK Electronic Collection of Papers Faculty of Civil Engineering*, 11, 20–36.

Sinha, P.C. (2002). *Encyclopedia of Disaster Management*. Anmol Publishers Private Limited, Delhi.

Sudhanshu, Joy, J., Singh, S.K., Kanga, S. (2021). Delineation of the evacuation route plan, relief camp, and prioritization using GIScience. *Advances in Environmental Research, an International Journal*, 10(1), 1–15.

Tomar, P., Singh, S.K., Kanga, S., Meraj, G., Kranjčić, N., Đurin, B., Pattanaik, A. (2021a). GIS-based urban flood risk assessment and management – A case study of Delhi National Capital Territory (NCT), India. *Sustainability*, 13(22), 12850. doi: 10.3390/su132212850.

Tomar, J.S., Kranjčić, N., Đurin, B., Kanga, S., Singh, S.K. (2021b). Forest fire hazards vulnerability and risk assessment in Sirmaur district forest of Himachal Pradesh (India): A geospatial approach. *ISPRS International Journal of Geo-Information*, 10(7), 447. doi: 10.3390/ijgi10070447.

Tomar, J.S., Kanga, S., Singh, S.K. (2021c). Giscience for forest fire modeling: New advances in wildland fires with management. *Plant Archives*, 21(1), 473–481. doi: 10.51470/PLANTARCHIVES.2021.v21.no1.067.

Tomaszewski, B. (2014). *Geographic Information System (GIS) for Disaster Management*. CRC Press, Boca Raton.

Tomaszewski, B., Judex, M., Szarzynski, J., Radestock, C., Wirkus, L. (2015). Geographic information systems for disaster response: A review. *Journal of Homeland Security and Emergency Management*, 12(3). doi: 10.1515/jhsem-2014-0082.

UNHCR (2017). Key concepts on climate change and disaster displacement. UNHCR Report, London.

United Nations (2018). *World Urbanization Prospects 2018*. United Nations Publications, London.

Wannous, C. and Velasquez, G. (2017). United Nations Office for Disaster Risk Reduction (UNISDR)-UNISDR's contribution to science and technology for disaster risk reduction and the role of the International Consortium on Landslides (ICL). In *4th World Landslide Forum*, 109–115. Ljubljana.

Wisner, B., Blaikie, P.M., Blaikie, P., Cannon, T., Davis, I. (2004). *At Risk: Natural Hazards, People's Vulnerability and Disasters*. Psychology Press, London [Online]. Available at: https://www.preventionweb.net.

Wisner, B., Gaillard, J.C., Kelman, I. (2012). *Handbook of Hazards and Disaster Risk Reduction*. Routledge, Abingdon.

Yadav, R.K. (2012). *Encyclopaedia of Disaster and Hazard Management*, volume 10. Arts & Science Academic Publishing, New Delhi.

Web details

BYJU'S: https://byjus.com/biology/ozone-layer-depletion.

Conserve Energy Future: www.conserve-energy-future.com.

Course Hero: https://www.coursehero.com/tutors-problems/Earth-Science/61564078-Essay.

eGyanKosh: https://egyankosh.ac.in/bitstream/123456789/26502/4/Unit-1.pdf.

Geology Science: https://geologyscience.com/natural-hazards/landslides/landslides/?amp.

Government of India: https://ndma.gov.in/Man-made-Hazards/Nuclear.

Government of India: https://ndma.gov.in/Man-made-Hazards/Nuclear/Dos-Donts.

Government of India: https://nidm.gov.in/PDF/Disaster-hymet.pdf.

KNM Sonamura: http://knmsonamura.ac.in/old-qp/S-Skill/S-Skill-Disaster-Management-QA.pdf.

Medium: https://medium.com/counterarts/90-of-the-worlds-earthquakes-occur-here-4ccce17e7d1.

Model Question Bank: http://knmsonamura.ac.in/old-qp/S-Skill/S-Skill-Disaster-Management-QA.pdf.

Nuclear Energy Agency: https://www.oecd-nea.org/jcms.

Quora: https://www.quora.com/What-is-seismograph-How-does-it-operate.

Research Gate: https://www.researchgate.net/post/What_are_the_main_causes_for_depleting_ozone_layer.

Studocu: https://www.studocu.com/in/document/indira-gandhi-national-open-university/understanding-natural-disasters/unit-11-understanding-natural-resources-and/23729539.

Tired Earth: https://tiredearth.com/articles/acid-rain-causes-effects-and-solutions.

Wikiwand: https://www.wikiwand.com/en/Earthquake_zones_of_India.

Index

G, H, I

L, M, N

O, P

R

S

T, U

V, W

Other titles from

iSTE

in

Ecological Science

2024

FLEURENCE Joël
Aquatic Biotechnologies: From Genetic Engineering to Enzymatic or Fermentation Engineering

MATAGNE Patrick
Birth of Scientific Ecology: Eugenius Warming (1841–1924)

2023

GAID Kader
Drinking Water Treatment 1: Water Quality and Clarification
Drinking Water Treatment 2: Chemical and Physical Elimination of Organic Substances and Particles
Drinking Water Treatment 3: Organic and Mineral Micropolluants
Drinking Water Treatment 4: Membranes Applied to Drinking Water and Desalination
Drinking Water Treatment 5: Calco-carbonic Equilibrium and Disinfection

2022

AMIARD Jean-Claude
Marine Radioecology
(Radioactive Risk Set – Volume 6)

GODET Laurent, DUFOUR Simon, ROLLET Anne-Julia
The Baseline Concept in Biodiversity Conservation: Being Nostalgic or Not in the Anthropocene Era

GRISON Claude, CASES Lucie, LE MOIGNE Mailys,
HOSSAERT-MCKEY Martine
Photovoltaism, Agriculture and Ecology: From Agrivoltaism to Ecovoltaism

LÉVÊQUE Christian
Biodiversity Erosion: Issues and Questions

2021

AMIARD Jean-Claude
Disarmament and Decommissioning in the Nuclear Domain
(Radioactive Risk Set – Volume 4)
Management of Radioactive Waste
(Radioactive Risk Set – Volume 5)

FLEURENCE Joël
Microalgae: From Future Food to Cellular Factory

JOLY Fernand, BOURRIÉ Guilhem
Mankind and Deserts 1: Deserts, Aridity, Exploration and Conquests
Mankind and Deserts 2: Water and Salts
Mankind and Deserts 3: Wind in Deserts and Civilizations

PIOCH Sylvain, SOUCHE Jean-Claude
Eco-design of Marine Infrastructures: Towards Ecologically-informed Coastal and Ocean Development

2020

BRUSLÉ Jacques, QUIGNARD Jean-Pierre
Fish Behavior 1: Eco-ethology
Fish Behavior 2: Ethophysiology

ROSSIGNOL Jean-Yves
Climatic Impact of Activities: Methodological Guide for Analysis and Action

2019

AMIARD Jean-Claude
Industrial and Medical Nuclear Accidents: Environmental, Ecological, Health and Socio-economic Consequences
(Radioactive Risk Set – Volume 2)
Nuclear Accidents: Prevention and Management of an Accidental Crisis
(Radioactive Risk Set – Volume 3)

BOULEAU Gabrielle
Politicization of Ecological Issues: From Environmental Forms to Environmental Motives

GIRAULT Yves
UNESCO Global Geoparks: Tension Between Territorial Development and Heritage Enhancement

KARA Mohamed Hichem, QUIGNARD Jean-Pierre
Fishes in Lagoons and Estuaries in the Mediterranean 2: Sedentary Fish
Fishes in Lagoons and Estuaries in the Mediterranean 3A: Migratory Fish
Fishes in Lagoons and Estuaries in the Mediterranean 3B: Migratory Fish

OUVRARD Benjamin, STENGER Anne
Incentives and Environmental Policies: From Theory to Empirical Novelties

2018

AMIARD Jean-Claude
Military Nuclear Accidents: Environmental, Ecological, Health and Socio-economic Consequences
(Radioactive Risk Set – Volume 1)

FLIPO Fabrice
The Coming Authoritarian Ecology

GUILLOUX Bleuenn
Marine Genetic Resources, R&D and the Law 1: Complex Objects of Use

KARA Mohamed Hichem, QUIGNARD Jean-Pierre
Fishes in Lagoons and Estuaries in the Mediterranean 1: Diversity, Bioecology and Exploitation

2016

BAGNÈRES Anne-Geneviève, HOSSAERT-MCKEY Martine
Chemical Ecology

2014

DE LARMINAT Philippe
Climate Change: Identification and Projections

Printed and bound by CPI Group (UK) Ltd, Croydon, CR0 4YY
16/09/2026
14963089-0004